Electron Capture
Negative Ion Mass Spectra of
Environmental Contaminants
and Related Compounds

Electron Capture Negative Ion Mass Spectra of Environmental Contaminants and Related Compounds

Elizabeth A. Stemmler
Ronald A. Hites

VCH

Elizabeth A. Stemmler
Ronald A. Hites
School of Public and Environmental Affairs
and Department of Chemistry
Indiana University
Bloomington, Indiana 47405

Library of Congress Cataloging-in-Publication Data

Stemmler, Elizabeth A. (Elizabeth Ann), 1960-
  Electron capture negative ion mass spectra of environmental
contaminants and related compounds.

  Includes index.
  1. Pollutants — Analysis. 2. Pollutants — Spectra.
3. Anions — Spectra. 4. Mass spectrometry.
I. Hites, R. A.   II. Title.
TD193.S74   1988   628.5   88-168
ISBN 0-89573-708-6

Printed in the United States of America.

ISBN 0-89573-708-6 VCH Publishers
ISBN 3-527-26904-5 VCH Verlagsgesellschaft

Distributed in North America by:

VCH Publishers, Inc.
220 East 23rd Street, Suite 909
New York, New York 10010

Distributed Worldwide by:

VCH Verlagsgesellschaft mbH
P.O. Box 1260/1280
D-6940 Weinheim
Federal Republic of Germany

PREFACE

The chemical ionization source of a mass spectrometer can be used to produce negative ions by electron capture reactions (electron capture negative ion mass spectrometry - ECNIMS) using a nonreactive enhancement gas such as methane or argon. Negative ions are efficiently and selectively formed by reactions of electrons with electronegative molecules. Since many environmental contaminants form negative ions in this manner, the electron capture detector is commonly used to measure these compounds. Because electron-molecule reactions proceed more rapidly than ion-molecule reactions, lower detection limits for certain compounds are observed when using ECNIMS instead of positive ionization mass spectrometry. ECNIMS has the additional advantage of being highly selective, permitting specific compound detection in complex matrices.

Although negative ions have been studied since the beginning of this century, they have not received the attention given to positive ions. In 1968, Ralph Dougherty showed that the reproducibility and intensity of negative ion mass spectra could be increased by the addition of a non-reactive enhancement gas, such as N_2, which served to increase the secondary electron current and provide a distribution of electron energies. The wide availability of chemical ionization sources, which limit gas conductance so that ion source pressures of around 1 torr can be attained, has made gas enhancement a popular method for the generation of negative ions. In this pressure range, the enhancement gas serves as a source of secondary electrons and as a medium for the thermalization of electrons and the stabilization of negative ions. Enhancement gases such as hydrocarbons and noble gases have been used. In addition, reactive reagent gases, such as oxygen, have been used to generate negative ions by ion-molecule reactions.

Because a chemical ionization source can be used to generate negative ions by either electron capture or ion-molecule reactions, there is considerable confusion regarding the proper name (and acronym) for this technique. The terms negative chemical ionization (NCI), negative ion chemical ionization (NICI), and electron capture-negative chemical ionization (EC-NCI) have all been used in the past. Electron capture negative ion (ECNI) mass spectrometry will be used here because it clearly indicates that electron capture, not ion-molecule, reactions were used to generate negative ions.

Other advances in negative ion mass spectrometry include the introduction of conversion dynode electron multipliers and pulsed positive ion negative ion chemical ionization mass spectrometry. The conversion dynode electron multiplier uses a conventional positive ion electron multiplier; however, the negative ions are converted to positive ions before they enter the multiplier. The negative to positive ion conversion is accomplished by accelerating the negative ions into a conversion dynode held at high positive potential. The negative ions cause sputtering of metal atoms (which may lose electrons and become positive ions), fragment and produce some positive ions, or undergo charge stripping to produce a neutral which may lose an electron to form a positive ion. This conversion dynode multiplier has eliminated the need to electrically float the multiplier at a high positive potential, reduced the noise due to stray electrons entering the multiplier, and improved the detection of the low energy negative ions encountered in quadrupoles.

This book contains the ECNI mass spectra of 361 compounds measured in our laboratory here at Indiana University. These compounds were selected to include the major classes of environmental contaminents such as those found on the U. S. Environmental Protection Agency's priority pollutant list. Compound classes of environmental importance include halogenated benzenes and phenols; nitro-benzenes and the related dinitro herbicides; polycyclic aromatic hydrocarbons; halogenated biphenyls, dioxins, and dibenzofurans; DDT derivatives; and hexachlorocyclopentadiene pesticides. The spectra, measured with the ion source at $100^{o}C$ and $250^{o}C$, are reported graphically. In addition, compound name, Chemical Abstract registry number, molecular weight, and the structure are given.

The mass spectra were measured on a Hewlett Packard 5985B GC/MS system. This is a differentially pumped quadrupole instrument system with an EI/CI source. A conversion dynode electron multiplier was used for negative ion detection. Modifications to the original instrument design included replacement of the jet separator with a modified transfer line to allow direct input of the capillary GC column effluent to the ion source, the addition of a direct inlet internal standard line, and the addition of a Baratron capacitance manometer.

Under ECNI conditions, methane (99.99%) was used as the reagent gas. To reduced background from the gas or regulator, a trap (3" x 1/4") of molecular sieve 5A and activated charcoal was placed in the gas line. Before entering the ion source, the reagent gas flows coaxially around the capillary column and through a transfer line heated to $280^{o}C$.

Samples were introduced through a DB-5, 30 m x 0.25 mm, fused silica column using helium as the carrier gas. The injection port was held at $280^{o}C$. Injections of 1 uL were made in the splitless mode. The ion volume sits inside a stainless steel block which can be heated to temperatures between $100^{o}C$ and $280^{o}C$. The temperature is monitored by a thermocouple in the ion source block.

When the GC oven was held at $100^{o}C$ and the ion source temperature at $250^{o}C$, typical operating conditions were an ion source pressure of 0.55 torr (measured with the capacitance manometer) and a methane flow of 15 mL/min. Under these conditions, the manifold pressure was about 3×10^{-4} torr. When the ion source temperature was lowered to $100^{o}C$, the ion source pressure droped to 0.49 torr.

The instrument was tuned using a 1000:1 mixture of perfluorotributylamine (PFTBA) and pentafluorobenzonitrile (PFBN); the target ions were m/z 633 and 452 from PFTBA and m/z 193 from PFBN. Peak widths at half height for the m/z 633 and 193 ions were set at 0.55 amu. Tuning required only a short pulse of the PFTBA/PFBN mixture. Typical values for the ion source parameters were: electron energy (200 eV), emission current (300 uA), repeller (-4 to -8 V), drawout (40 to 60 V), ion focus (40 to 50 V), entrance lens (100 mV/amu). After tuning, the internal standard lines were evacuated for 0.5 hr to remove the tuning mixture (PFBN is particularly persistent). Background spectra were acquired, and a 100 pg injection of decafluorotriphenylphosphine (DFTPP) was made to check instrument performance. The m/z 442, 365, 275 and 167 ions were monitored by selected ion monitoring (SIM) using five 0.1 amu scans across each mass peak. These measurements were used to check the mass axis and to determine if the following relative ion abundance criteria were met: m/z 442 (> 60%) m/z 365 (> 10%), m/z 275 (< 10%), m/z 167 (< 20%). If the abundances were out of range, the ion focus and drawout lenses were used to adjust abundances.

ECNI mass spectra were measured in the full scan mode and were acquired using 1 or 2 A-to-D conversions per 0.125 amu per sec. This number was chosen, depending upon the mass range scanned, to obtain a scan speed of 1 scan/sec or less. The mass range scanned was typically from 10 amu to 50 amu above the molecular ion. All of the mass spectra reported here gave a base peak with an absolute intensity of at least 10^3 times the threshold value; thus, all peaks of 1% intensity or greater are reported.

We would like to thank the United States Department of Energy (Grant No. 80EV-10449) and the United States Environmental Protection Agency (Grant No. R808865) for the research support which made this book possible and Clark Porter, Ilora Basu, and Vicki Hites for clerical assistance.

CONTENTS

DINITRO HERBICIDE DERIVATIVES

NAME	CAS #	MW	PAGE

1,1,1-Trichloroethane
CAS No: 71-55-6
Formula: C$_2$H$_3$Cl$_3$ MW: 132

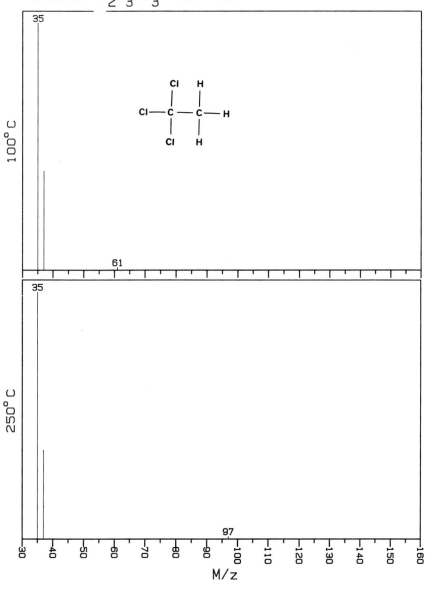

1, 1, 2-Trichloroethane
CAS No: 79-00-5
Formula: $C_2H_3Cl_3$ MW: 132

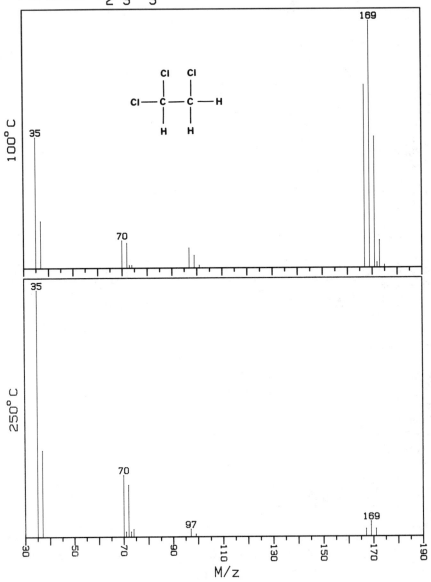

1,1,1,2-Tetrachloroethane
CAS No: 630-20-6
Formula: $C_2H_2Cl_4$ MW: 166

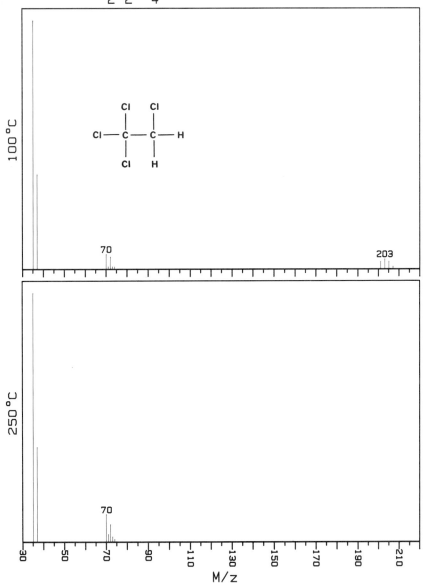

1, 1, 2, 2-Tetrachloroethane
CAS No: 79-34-5
Formula: $C_2H_2Cl_4$ MW: 166

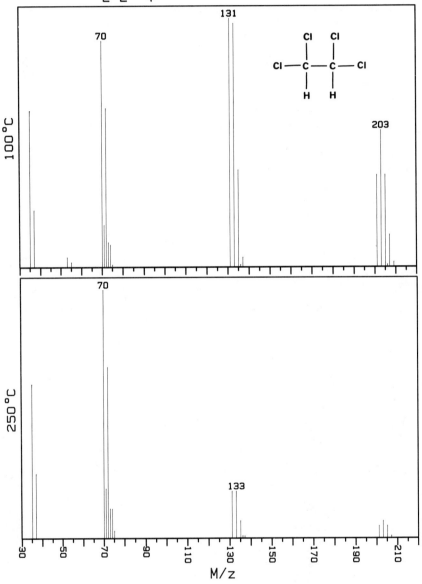

Pentachloroethane
CAS No: 76-01-7
Formula: C_2HCl_5 MW: 200

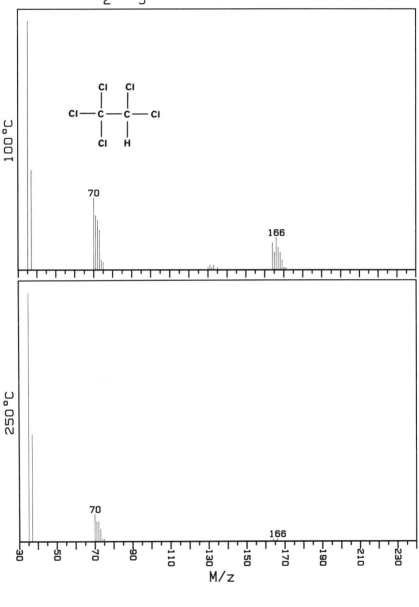

Hexachloroethane
CAS No: 67-72-1
Formula: C_2Cl_6 MW: 234

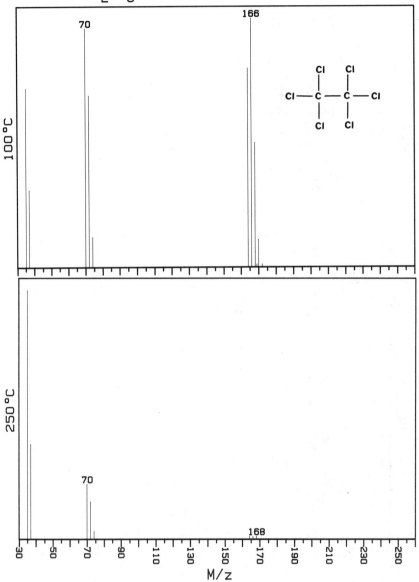

1,2,3-Trichloropropane
CAS No: 96-18-4
Formula: $C_3H_5Cl_3$ MW: 146

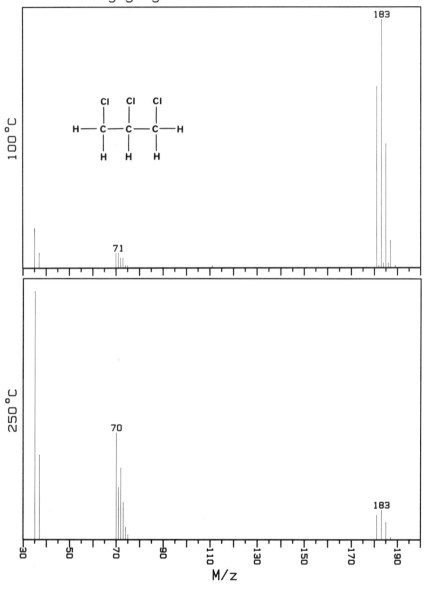

Dibromopropane
CAS No: 96-12-8
Formula: $C_3H_5Br_2Cl$ MW: 234

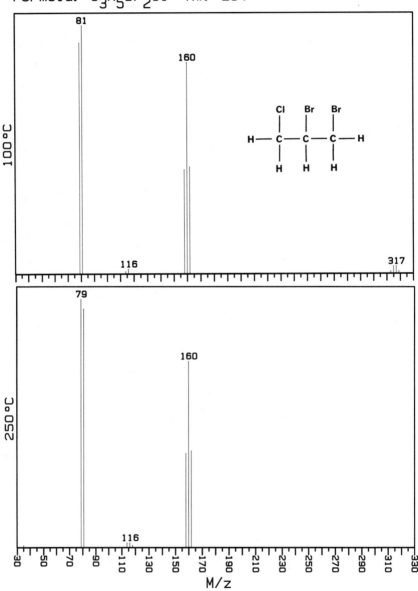

Octachloropropane
CAS No: 594-90-1
Formula: C$_3$Cl$_8$ MW: 316

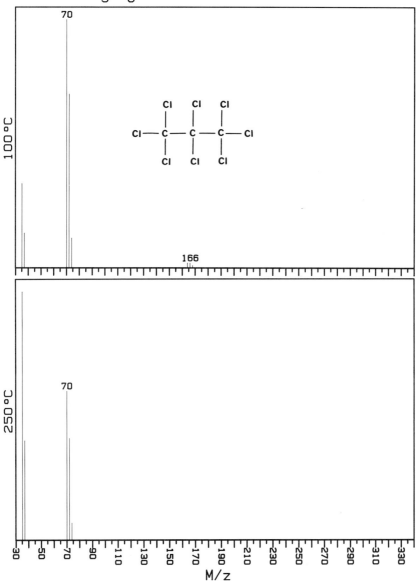

Trichloroethene
CAS No: 79-01-6
Formula: C_2HCl_3 MW: 130

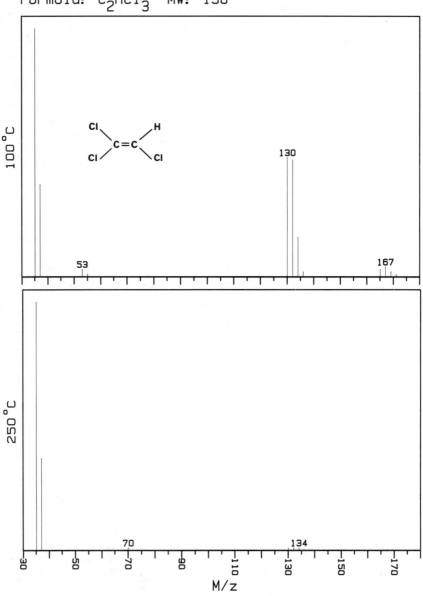

Tetrachloroethene
CAS No: 127-18-4
Formula: C_2Cl_4 MW: 164

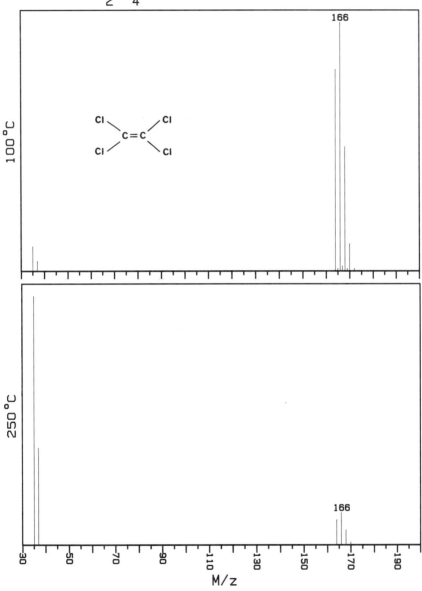

Tetrabromoethene
CAS No: 79-28-7
Formula: C_2Br_2 MW: 340

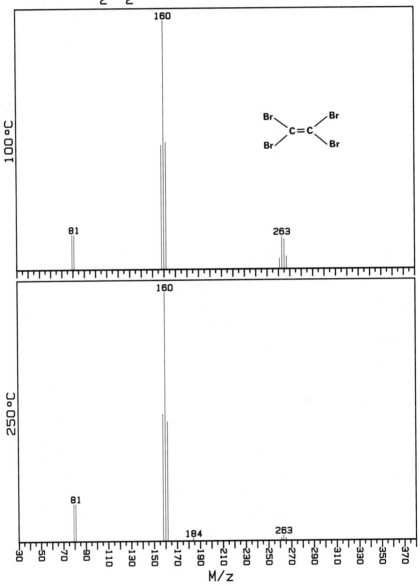

Tetraiodoethene
CAS No: 513-92-8
Formula: C_2I_4 MW: 532

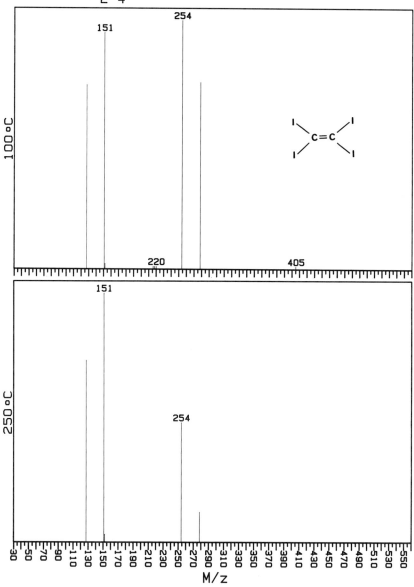

2,3-Dichlorohexafluoro-2-butene
CAS No: 303-04-8
Formula: $C_4Cl_2F_6$　MW: 232

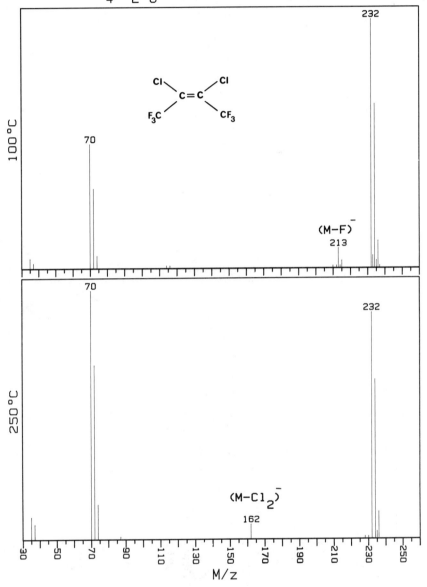

Hexachloropropene
CAS No: 1888-71-7
Formula: C_3Cl_6 MW: 246

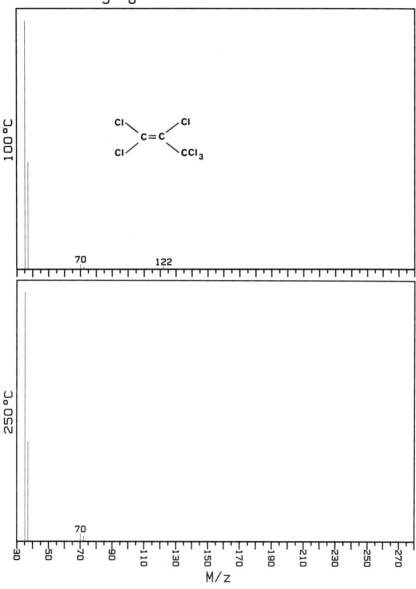

Hexachloro-1,3-butadiene
CAS No: 87-68-3
Formula: C_4Cl_6 MW: 258

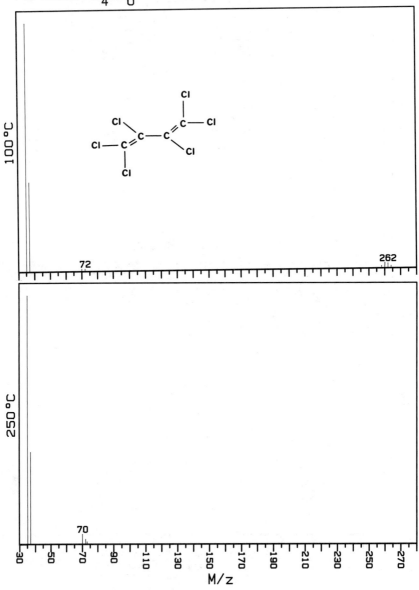

BHC, alpha isomer
CAS No: 319-84-6
Formula: $C_6H_6Cl_6$ MW: 288

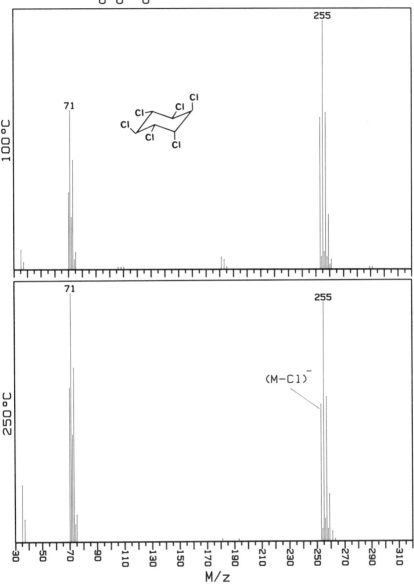

BHC, beta isomer
CAS No: 319-85-7
Formula: $C_6H_6Cl_6$ MW: 288

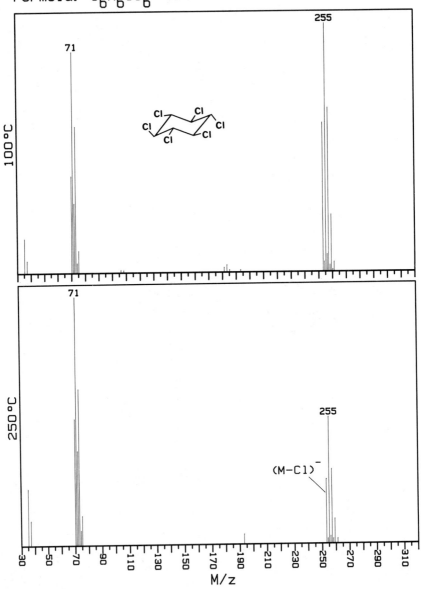

BHC, delta isomer
CAS No: 319-86-8
Formula: C₆H₆Cl₆   MW: 288

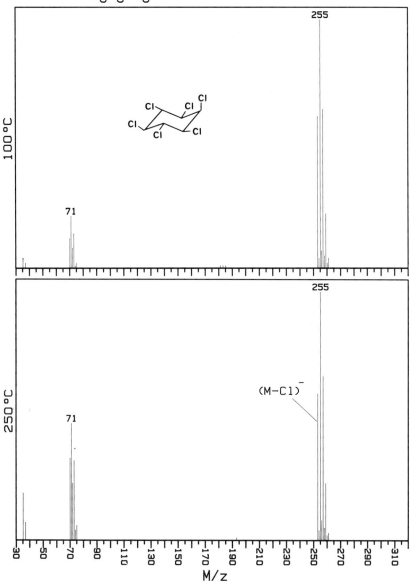

BHC, gamma isomer
CAS No: 58-89-9
Formula: $C_6H_6Cl_6$ MW: 288

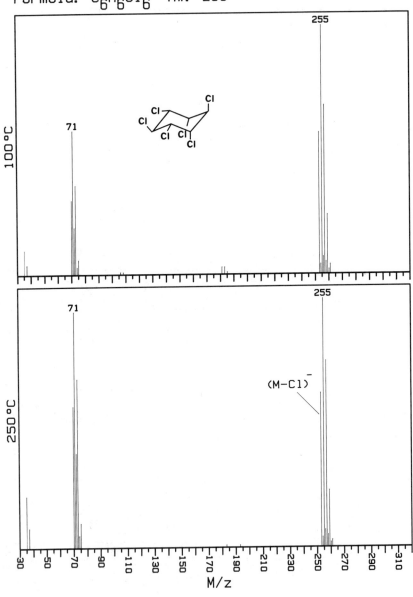

Hexachlorocyclopentadiene
CAS No: 77-47-4
Formula: C_5Cl_6, MW: 270

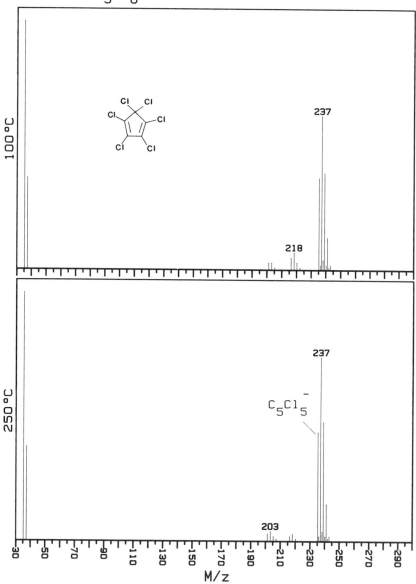

Hexachloro-3-cyclopentenone
CAS No: 15743-12-1
Formula: C_5Cl_6O MW: 286

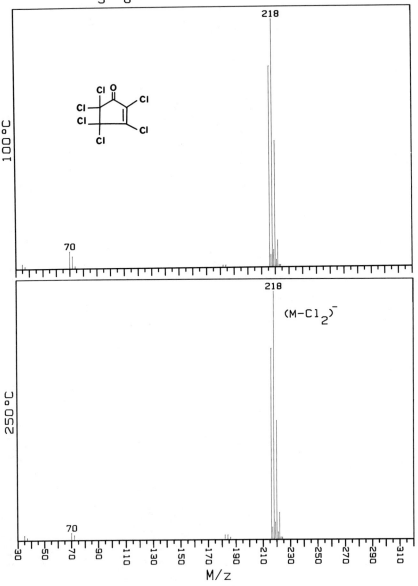

2,2,4,5-Tetrachlorocyclopentene-1,3-dione
CAS No: 15743-13-2
Formula: $C_5Cl_4O_2$ MW: 232

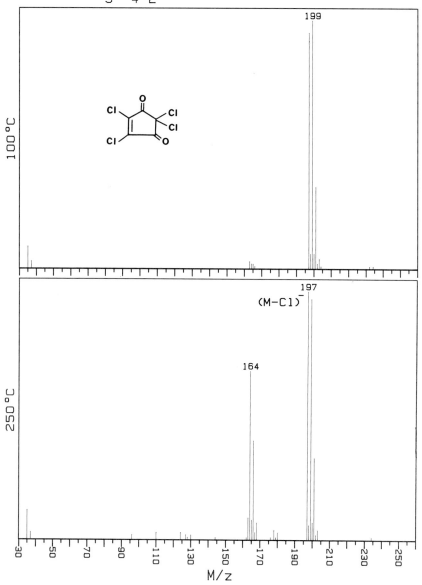

Nitrobenzene
CAS No: 98-95-3
Formula: $C_6H_5NO_2$ MW: 123

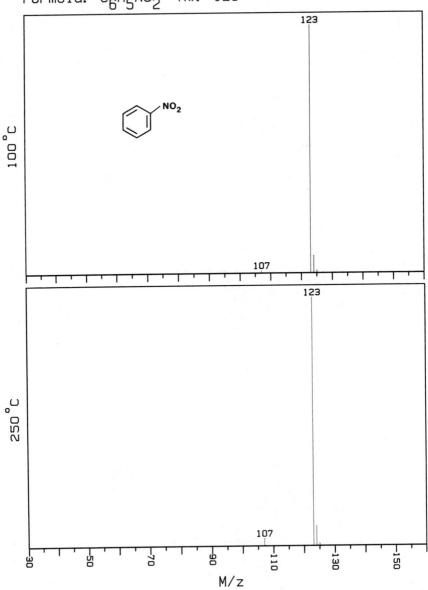

Chlorobenzene
CAS No: 108-90-7
Formula: C$_6$H$_5$NCl   MW: 112

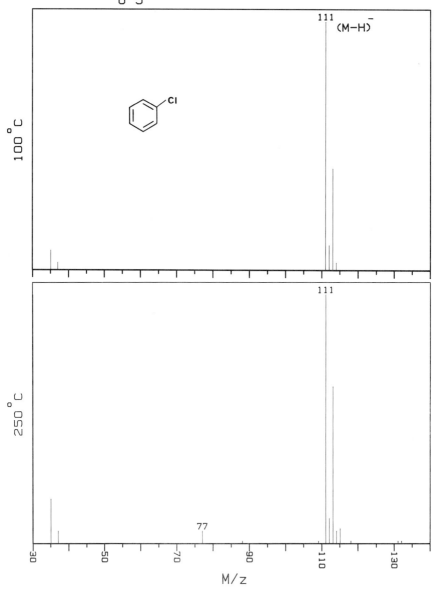

Bromobenzene
CAS No: 108-86-1
Formula: C_6H_5NBr MW: 156

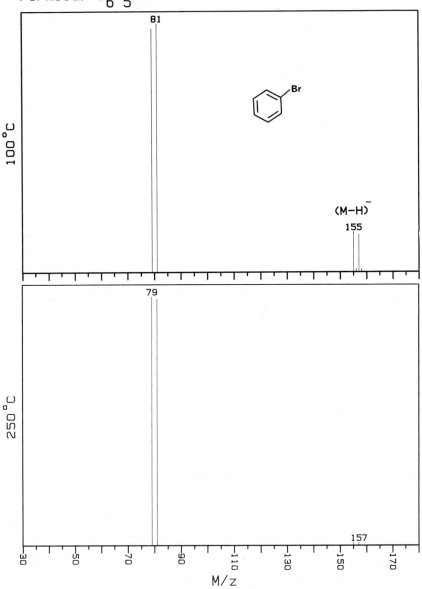

Iodobenzene
CAS No: 591-50-4
Formula: C₆H₅I   MW: 204

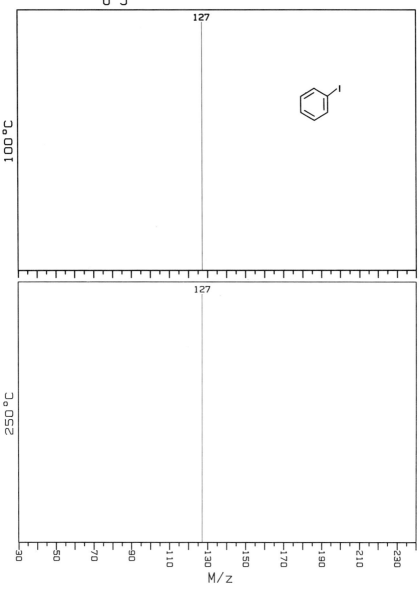

Benzonitrile
CAS No: 100-47-0
Formula: C_7H_5N MW: 103

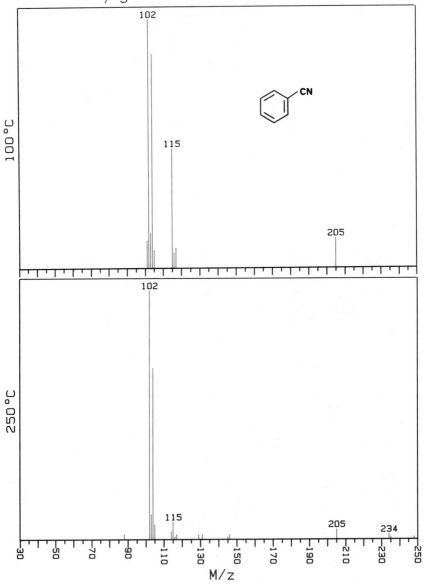

Phenol
CAS No: 108-95-2
Formula: C_6H_6O MW: 94

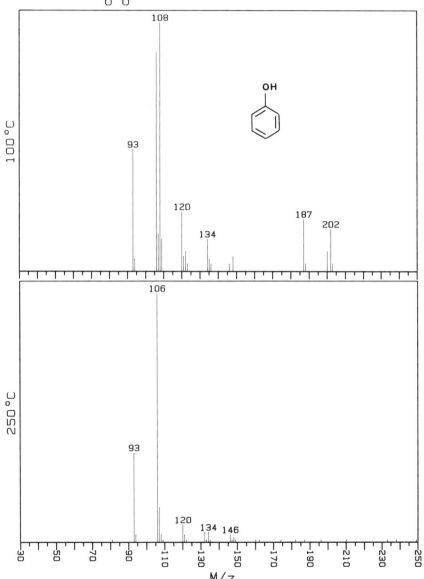

1,2-Dichlorobenzene
CAS No: 95-50-1
Formula: $C_6H_4Cl_2$ MW: 146

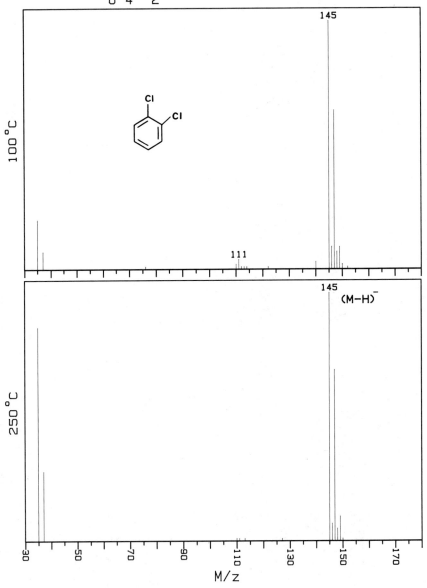

1,3-Dichlorobenzene
CAS No: 541-73-1
Formula: $C_6H_4Cl_2$ MW: 146

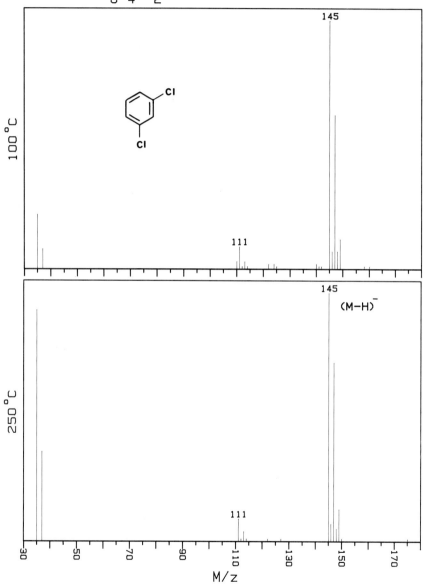

1,4-Dichlorobenzene
CAS No: 106-46-7
Formula: $C_6H_4Cl_2$ MW: 146

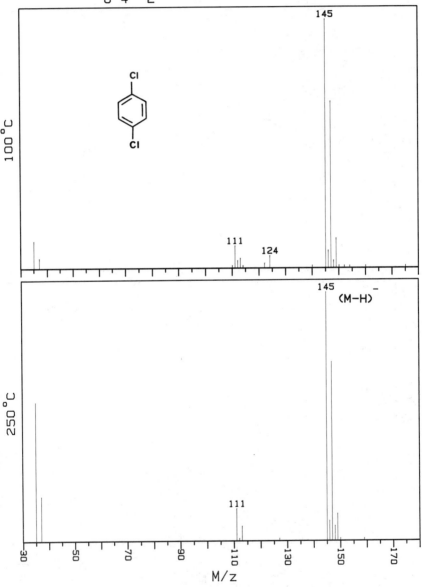

1,2,3-Trichlorobenzene
CAS No: 87-61-6
Formula: $C_6H_3Cl_3$ MW: 180

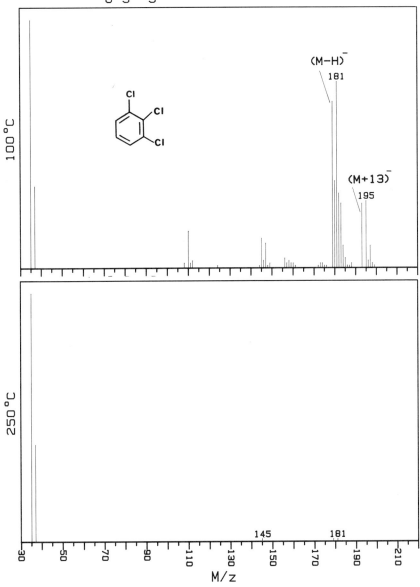

1,3,5-Trichlorobenzene
CAS No: 108-70-3
Formula: $C_6H_3Cl_3$ MW: 180

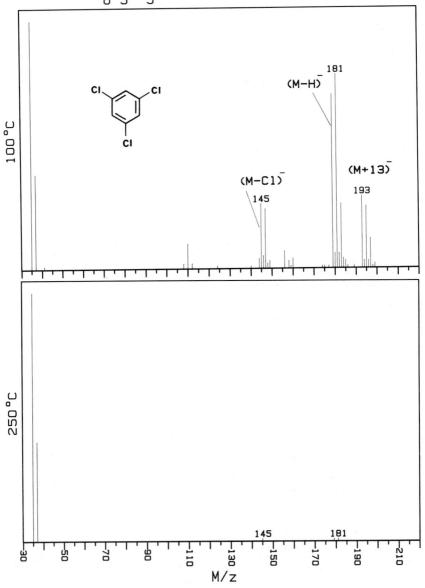

1, 2, 4-Trichlorobenzene
CAS No: 120-82-1
Formula: $C_6H_3Cl_3$ MW: 180

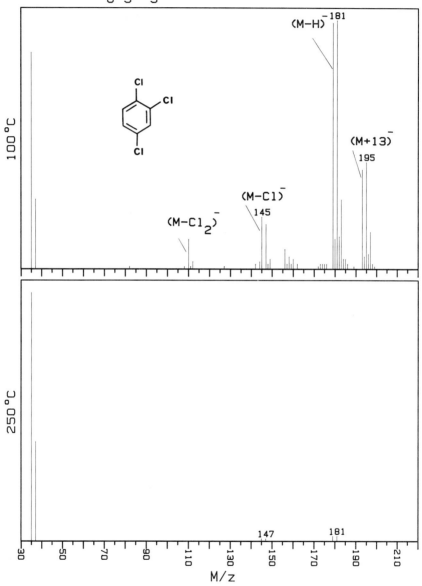

1, 2, 4, 5-Tetrachlorobenzene
CAS No: 95-94-3
Formula: $C_6H_2Cl_4$ MW: 214

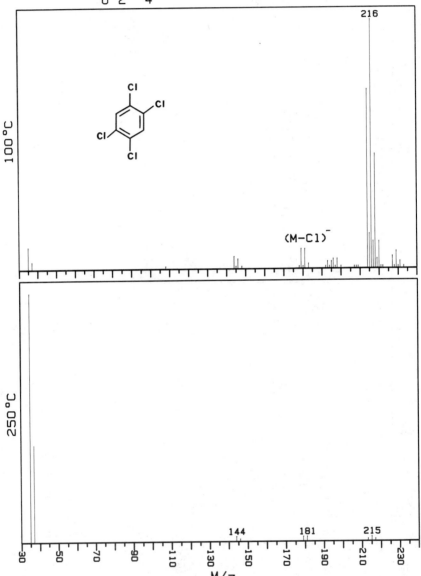

1, 2, 3, 4-Tetrachlorobenzene
CAS No: 634-66-2
Formula: $C_6H_2Cl_4$ MW: 214

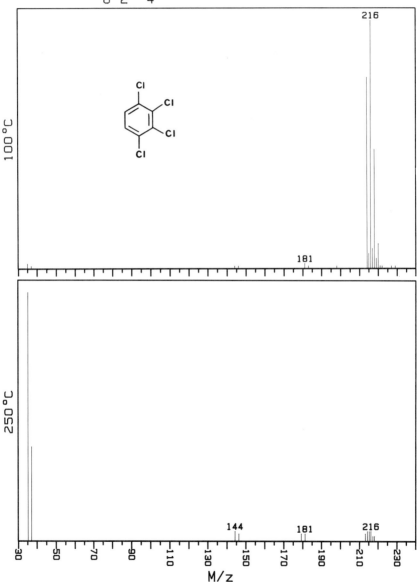

Pentachlorobenzene
CAS No: 608-93-5
Formula: C_6HCl_5 MW: 248

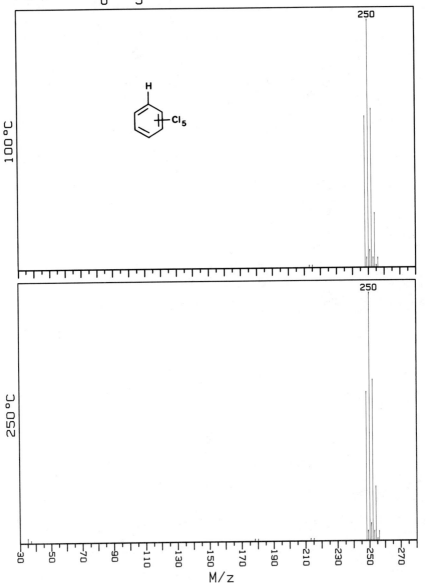

Hexachlorobenzene
CAS No: 118-74-1
Formula: C_6Cl_6 MW: 282

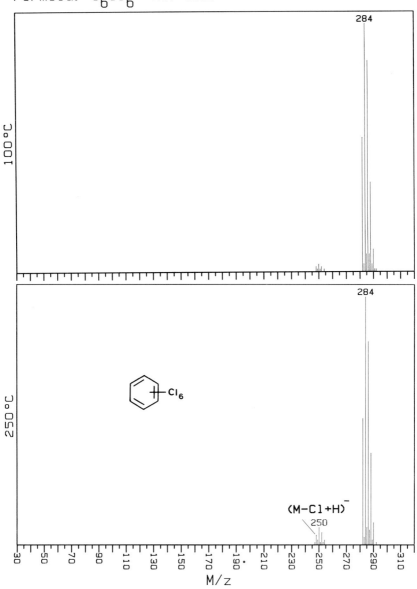

Octachlorostyrene
CAS No: 29082-74-4
Formula: C_8Cl_8 MW: 376

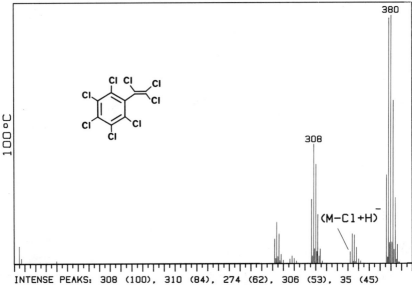

INTENSE PEAKS: 308 (100), 310 (84), 274 (62), 306 (53), 35 (45)

Octachlorostyrene
CAS No:
Formula: C_8Cl_8 MW: 376

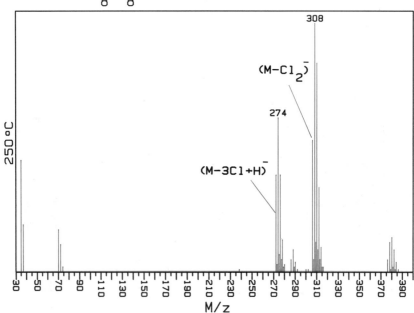

1,3,5-Tribromobenzene
CAS No: 626-39-1
Formula: $C_6H_3Br_3$ MW: 312

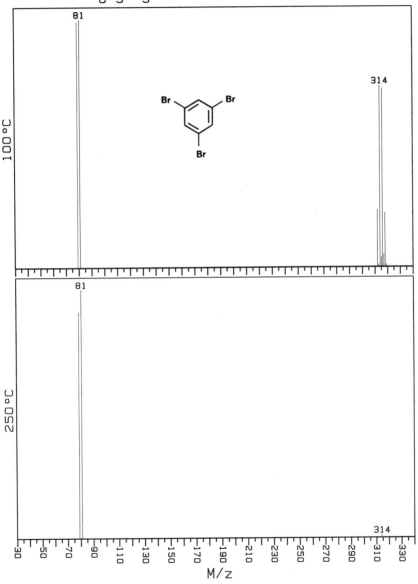

1,2,4-Tribromobenzene
CAS No: 615-54-3
Formula: $C_6H_3Br_3$ MW: 312

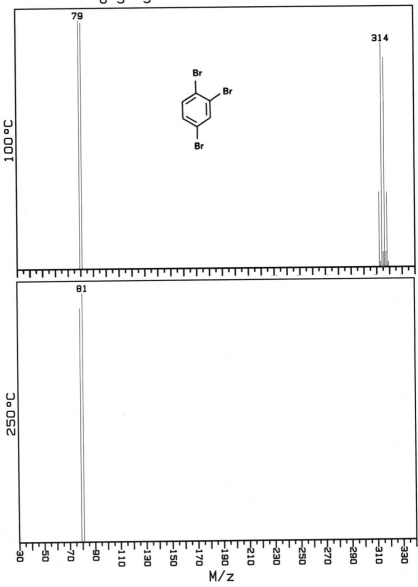

1, 2, 4, 5-Tetrabromobenzene
CAS No: 636-28-2
Formula: $C_6H_2Br_4$ MW: 390

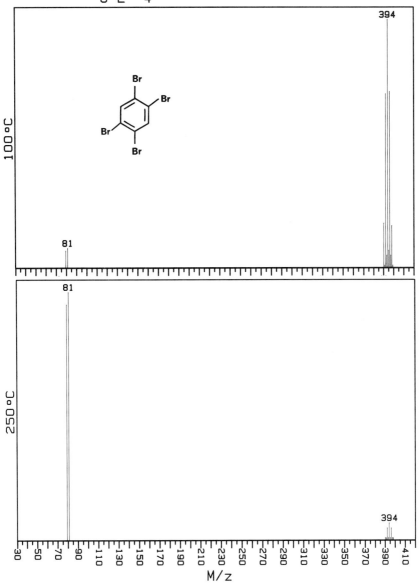

Hexabromobenzene
CAS No: 87-82-1
Formula: C_6Br_6 MW: 546

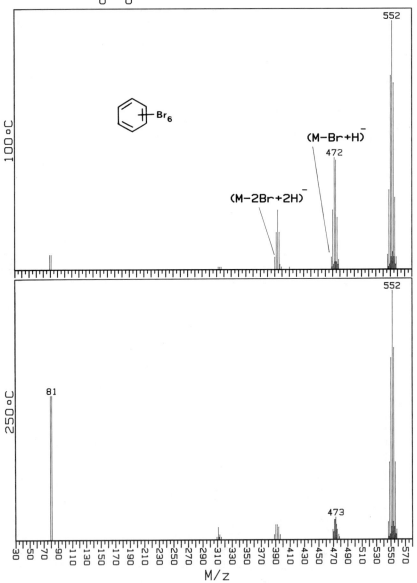

Hexafluorobenzene
CAS No: 392-56-3
Formula: C_6F_6 MW: 186

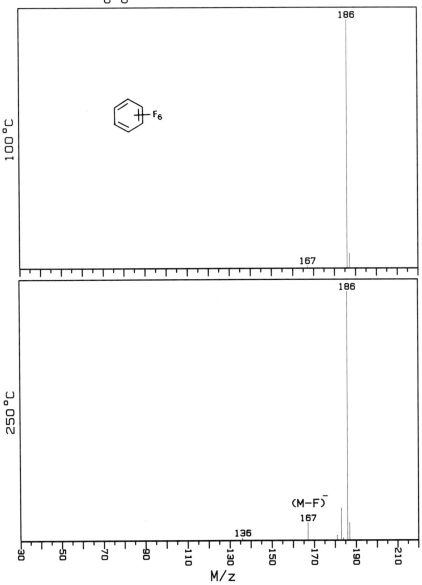

Chloropentafluorobenzene
CAS No: 344-07-0
Formula: C_6ClF_5 MW: 202

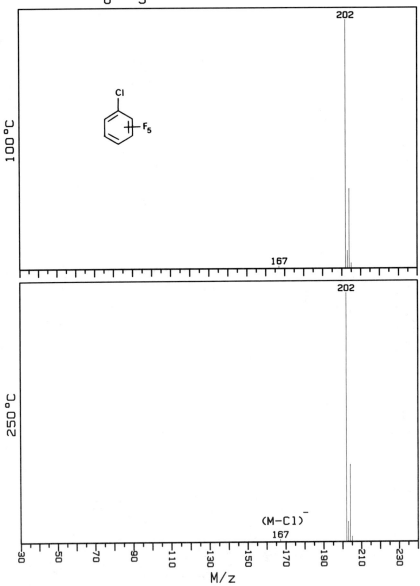

Bromopentafluorobenzene
CAS No: 344-04-7
Formula: C$_6$BrF$_5$, MW: 246

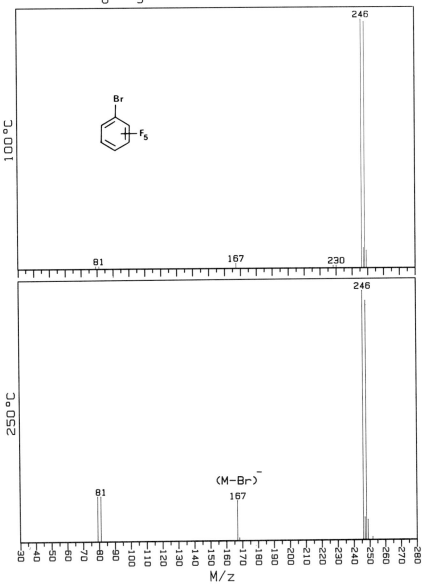

Iodopentafluorobenzene
CAS No: 827-15-6
Formula: C_6F_5I MW: 294

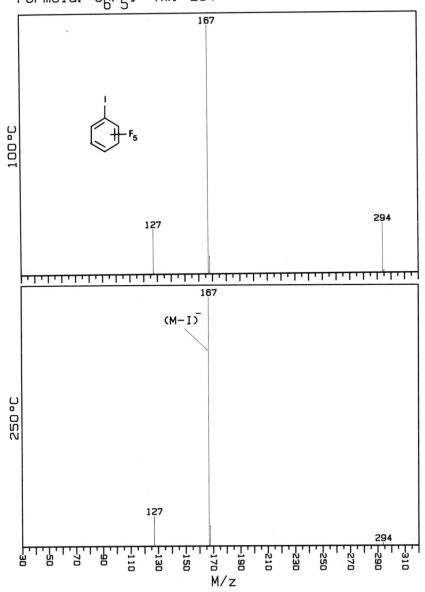

1-Chloro-2-nitrobenzene
CAS No: 88-73-3
Formula: $C_6H_4ClNO_2$ MW: 157

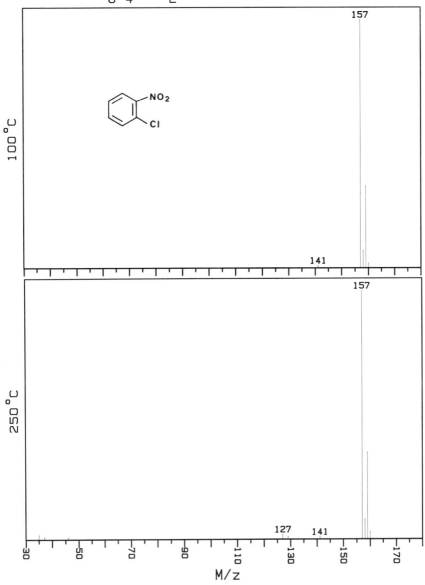

1-Chloro-3-nitrobenzene
CAS No: 121-73-3
Formula: $C_6H_4ClNO_2$ MW: 157

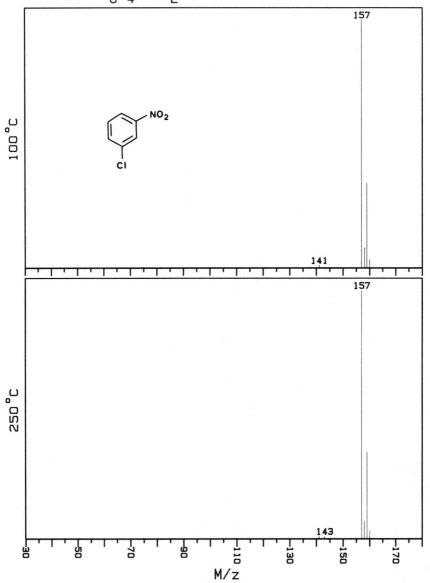

1-Chloro-4-nitrobenzene
CAS No: 100-00-5
Formula: $C_6H_4ClNO_2$ MW: 157

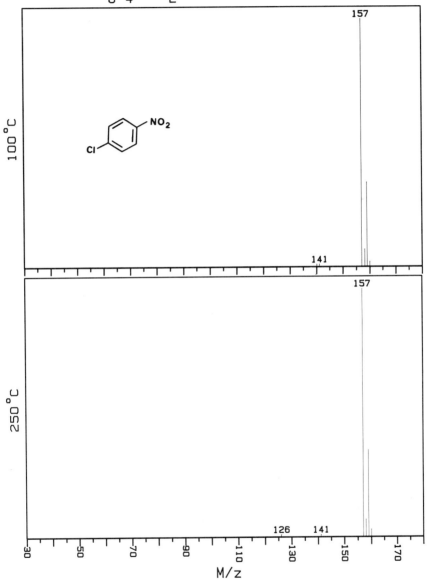

1, 2, 3-Trichloro-4-nitrobenzene
CAS No: 17700-09-3
Formula: $C_6H_2Cl_3NO_2$ MW: 225

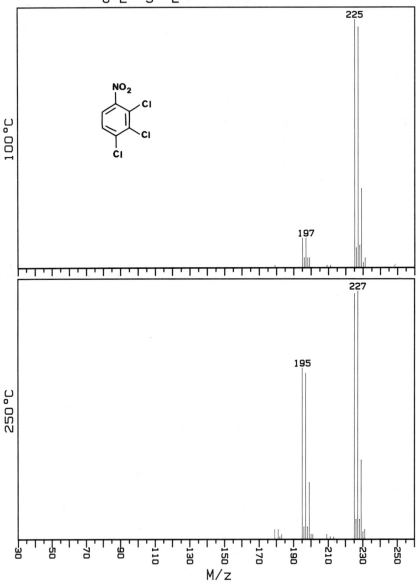

1, 2, 4-Trichloro-5-nitrobenzene
CAS No: 89-69-0
Formula: $C_6H_2Cl_3NO_2$ MW: 225

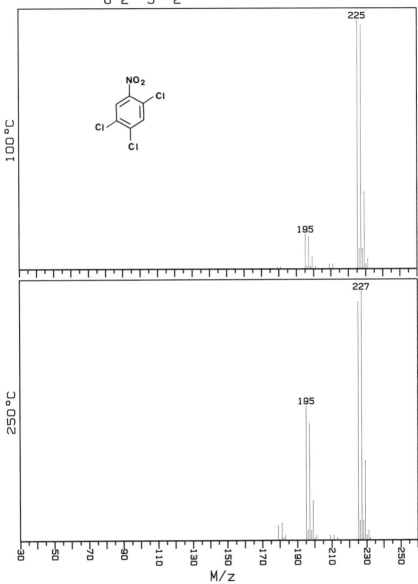

1,3,5-Trichloro-2-nitrobenzene
CAS No: 18708-70-8
Formula: $C_6H_2Cl_3NO_2$ MW: 225

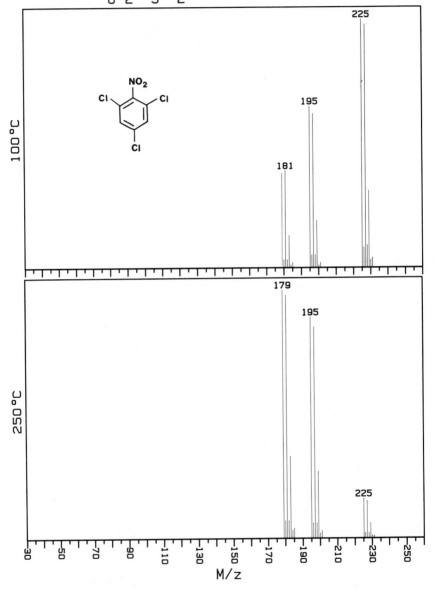

2,3,4,5-Tetrachloronitrobenzene
CAS No: 879-39-0
Formula: $C_6HCl_4NO_2$ MW: 259

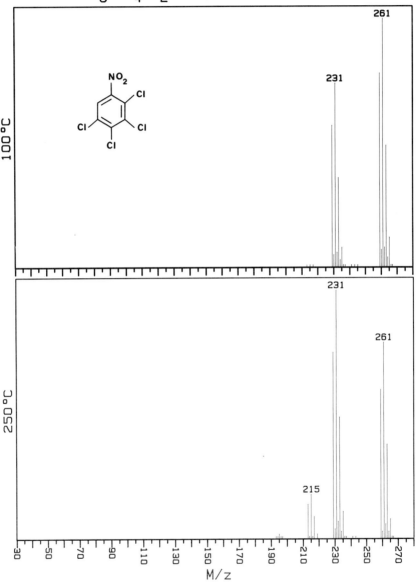

2,3,5,6-Tetrachloronitrobenzene
CAS No: 117-18-0
Formula: $C_6HCl_4NO_2$ MW: 259

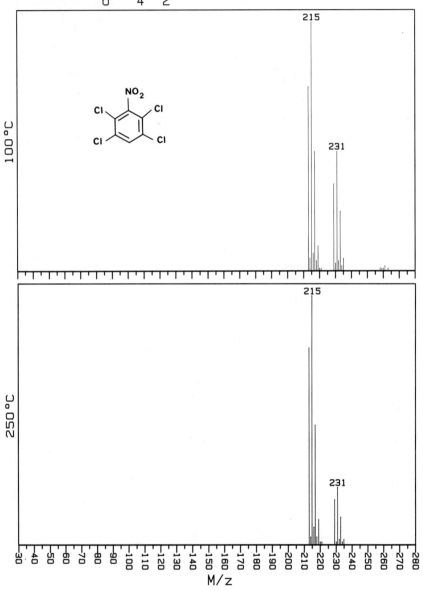

Pentachloronitrobenzene
CAS No: 82-68-8
Formula: $C_6Cl_5NO_2$ MW: 293

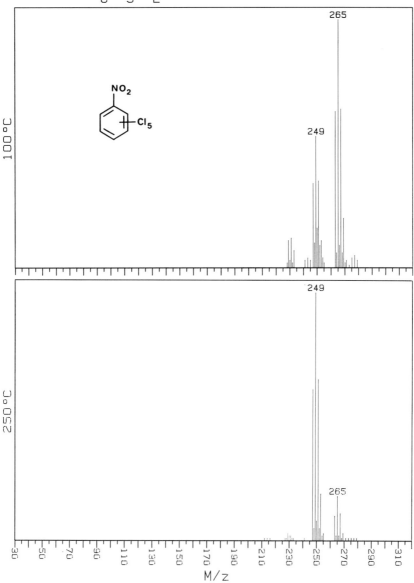

Pentachlorobenzonitrile
CAS No: 20925-85-3
Formula: C_7Cl_5N, MW: 273

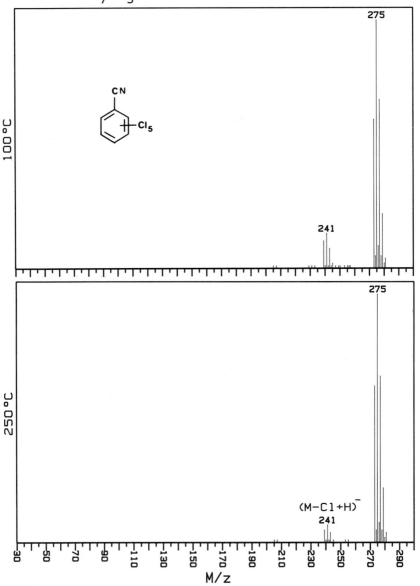

Chlorothalonil
CAS No: 1897-45-6
Formula: $C_8Cl_4N_2$ MW: 264

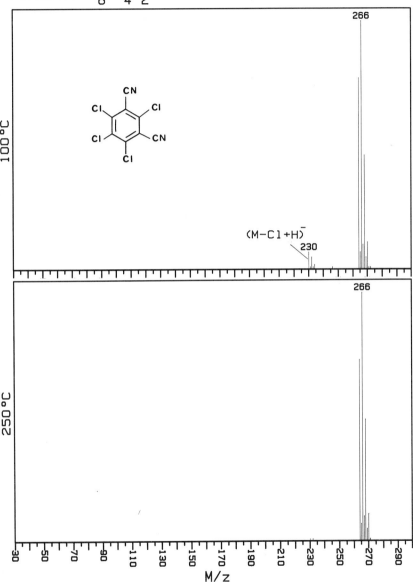

Pentafluorobenzonitrile
CAS No: 773-82-0
Formula: C_7F_5N, MW: 193

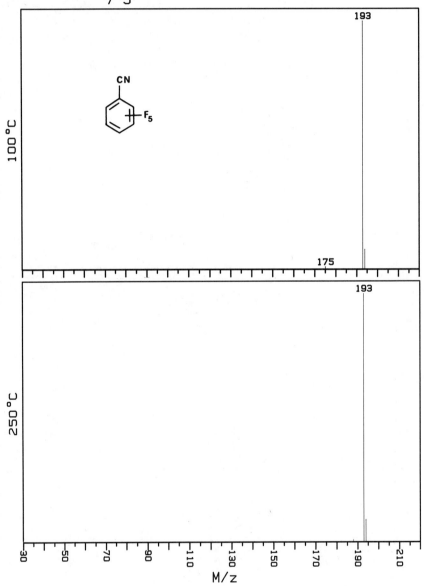

2, 3, 4-Trichlorophenol
CAS No: 15950-66-0
Formula: $C_6H_3Cl_3O$ MW: 196

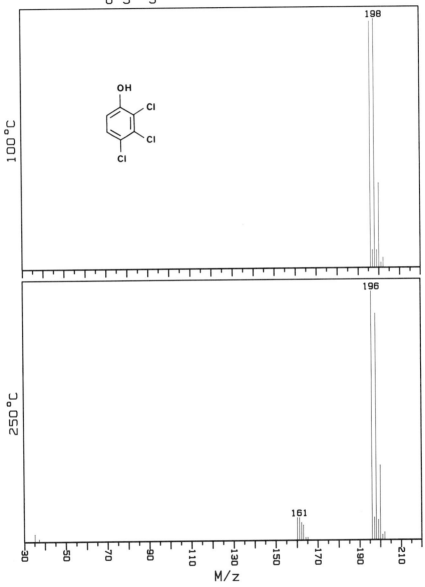

2, 3, 5-Trichlorophenol
CAS No: 933-78-8
Formula: $C_6H_3Cl_3O$ MW: 196

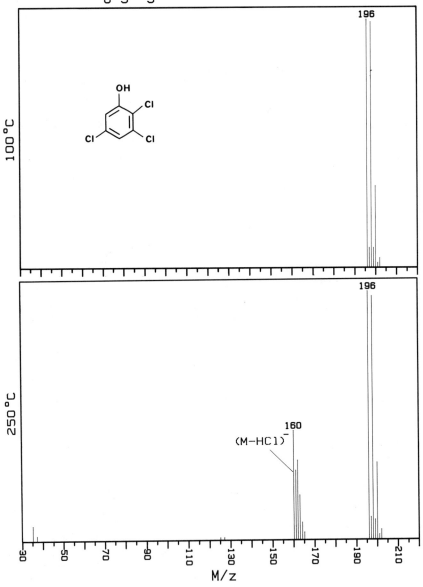

2,3,6-Trichlorophenol
CAS No: 933-75-5
Formula: $C_6H_3Cl_3O$ MW: 196

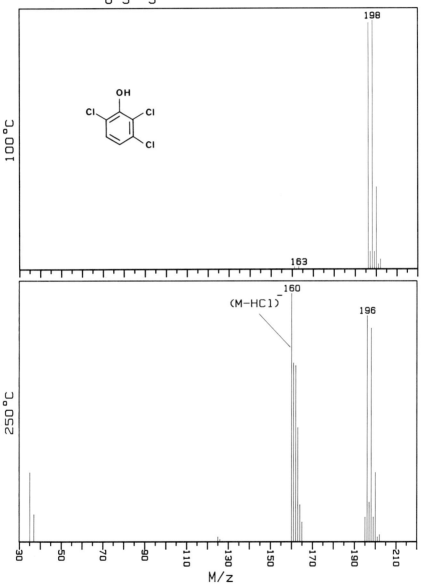

2, 4, 5-Trichlorophenol
CAS No: 95-95-4
Formula: $C_6H_4Cl_3O$ MW: 196

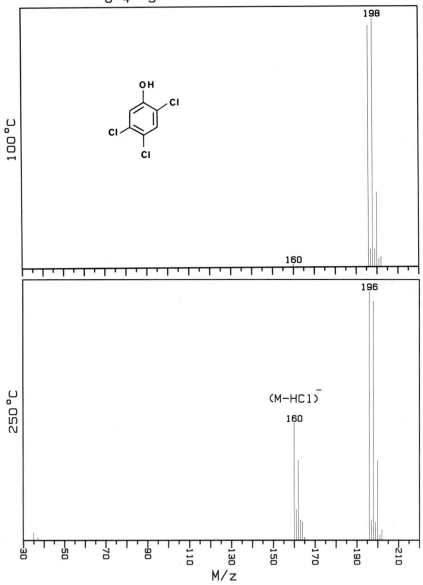

2,4,6-Trichlorophenol
CAS No: 88-06-2
Formula: $C_6H_3Cl_3O$ MW: 196

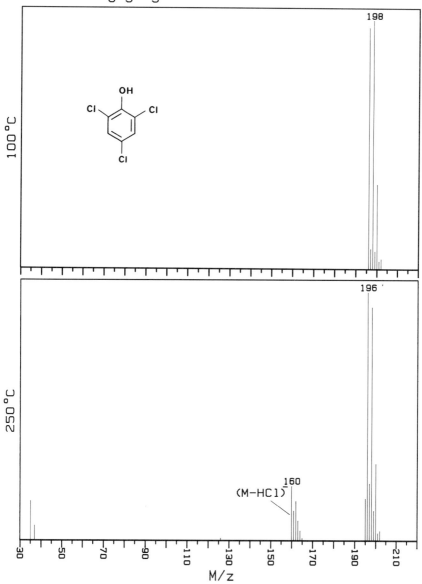

3, 4, 5-Trichlorophenol
CAS No: 609-19-8
Formula: $C_6H_3Cl_3O$ MW: 196

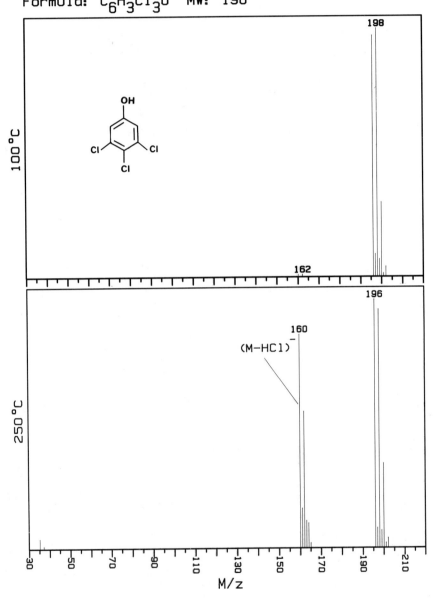

2,3,4,5-Tetrachlorophenol
CAS No: 4901-51-3
Formula: $C_6H_2Cl_4O$ MW: 230

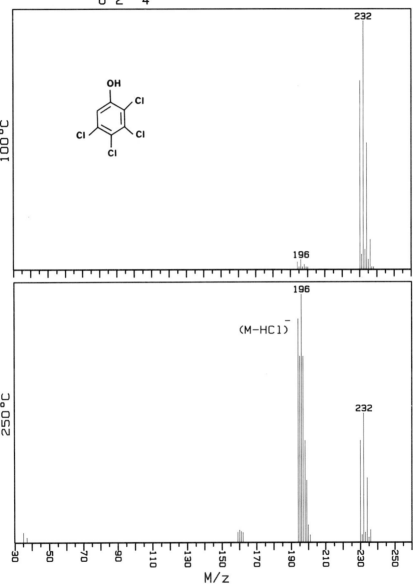

2, 3, 5, 6-Tetrachlorophenol
CAS No: 935-95-5
Formula: $C_6H_2Cl_4O$ MW: 230

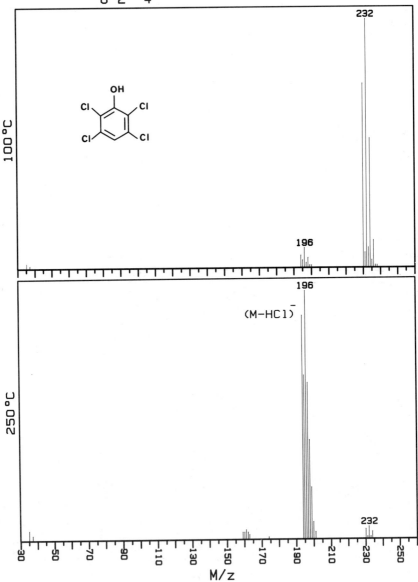

2,4,6-Tribromophenol
CAS No: 118-79-6
Formula: $C_6H_3Br_3O$ MW: 328

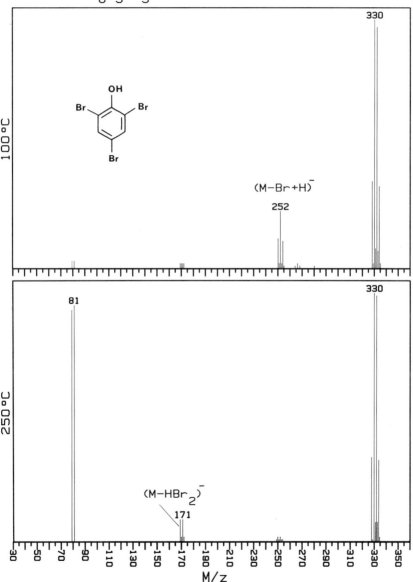

Pentachlorophenol
CAS No: 87-86-5
Formula: C_6HCl_5O MW: 264

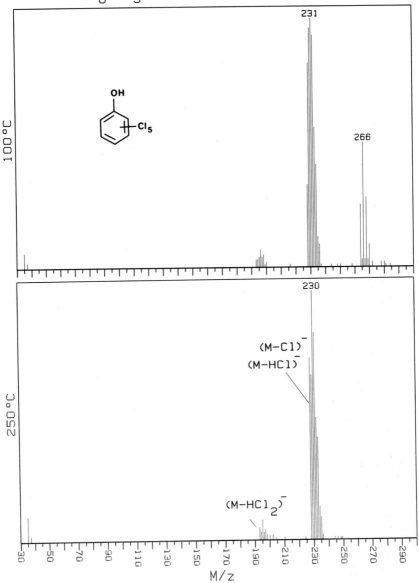

Pentabromophenol
CAS No: 608-71-9
Formula: C_6HBr_5O, MW: 484

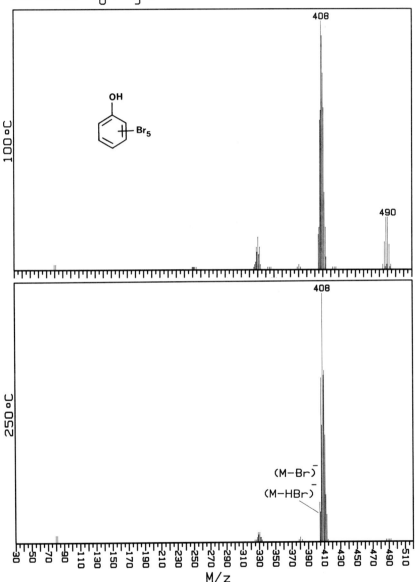

Pentafluorophenol
CAS No: 771-61-9
Formula: C_6HF_5O MW: 184

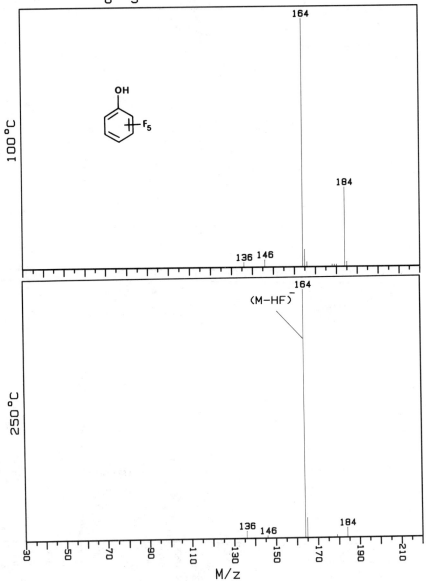

Pentachlorothiophenol
CAS No: 133-49-3
Formula: C$_6$HCl$_5$S   MW: 280

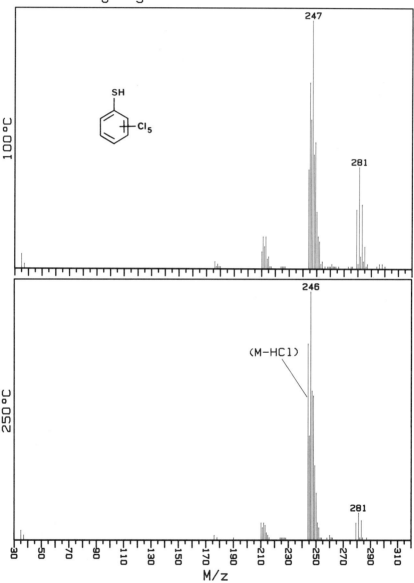

4-Bromo-2,5-dichlorophenol
CAS No: 1940-42-7
Formula: $C_6H_3BrCl_2O$ MW: 240

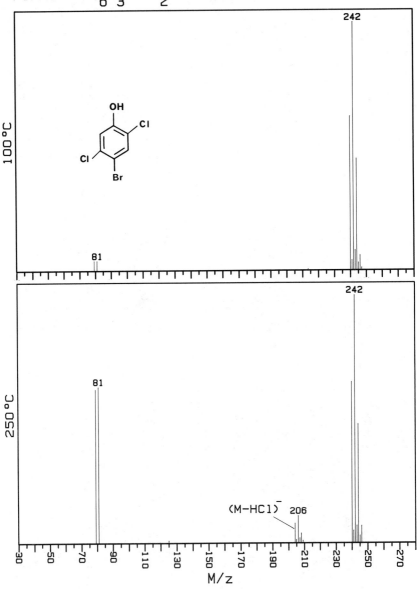

Bromoxynil
CAS No: 1689-84-5
Formula: $C_7H_3Br_2NO$ MW: 275

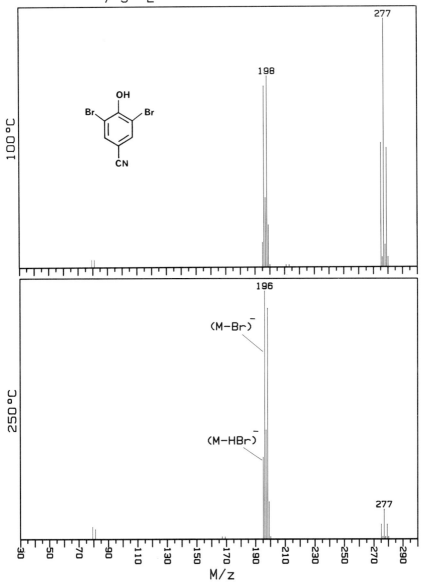

Ioxynil
CAS No: 1689-83-4
Formula: C$_7$H$_3$I$_2$NO   MW: 371

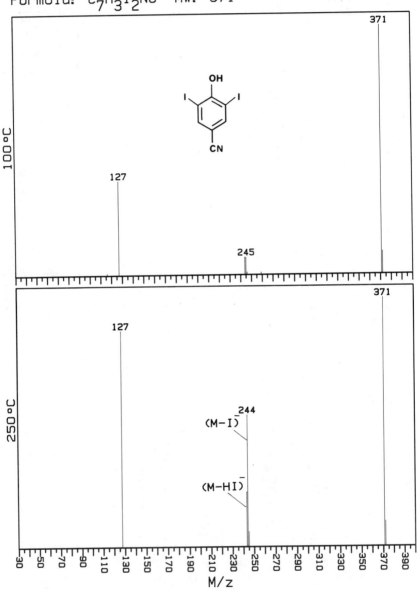

2,4,6-Trichloroanisole
CAS No: 50375-10-5
Formula: $C_7H_5Cl_3O$ MW: 210

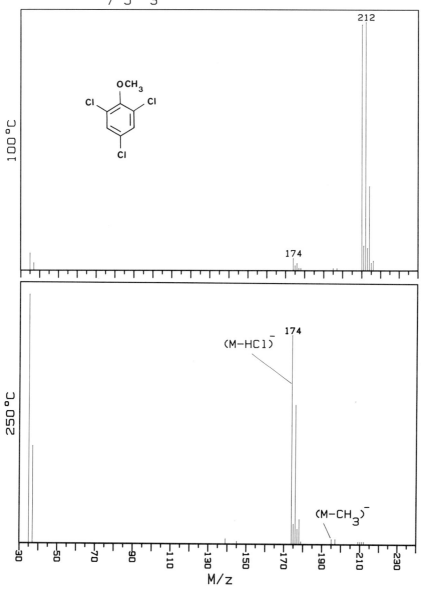

Pentachloroanisole
CAS No: 1825-21-4
Formula: $C_7H_3Cl_5O$ MW: 278

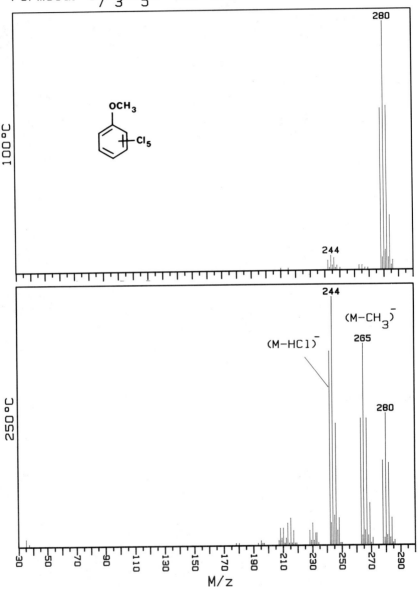

Pentafluoroanisole
CAS No: 389-40-2
Formula: C$_7$H$_3$F$_5$O   MW: 198

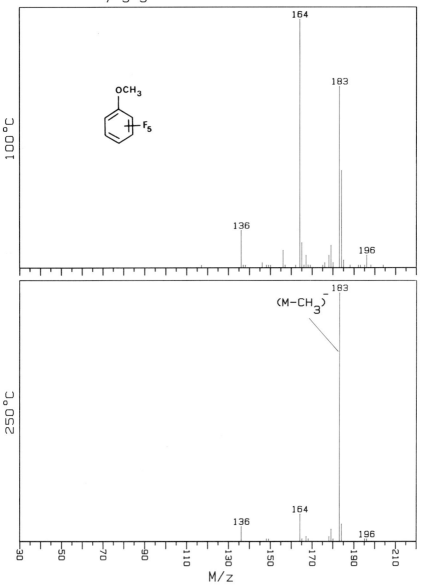

Chloroneb
CAS No: 2675-77-6
Formula: $C_8H_8Cl_2O_2$ MW: 206

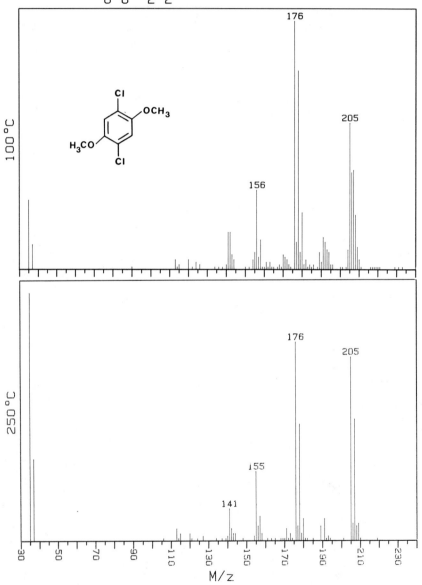

3, 4, 5-Trichloroaniline
CAS No: 634-91-3
Formula: $C_6H_4Cl_3N$ MW: 195

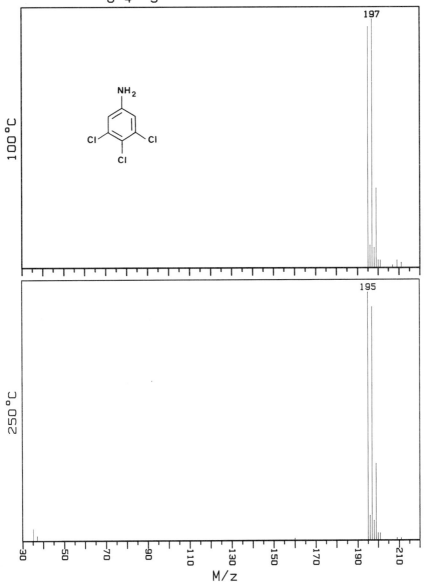

2, 4, 6-Tribromoaniline
CAS No: 147-82-0
Formula: $C_6H_4Br_3N$ MW: 327

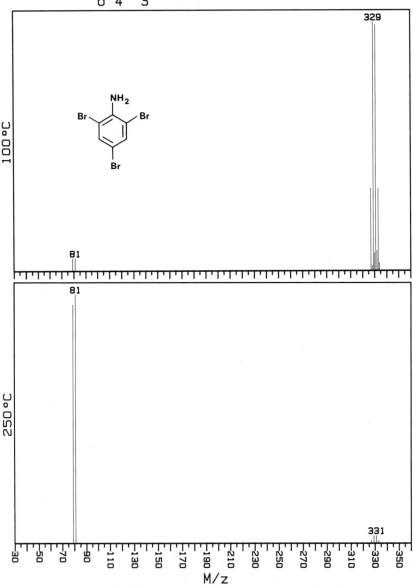

2, 3, 4, 5-Tetrachloroaniline
CAS No: 634-83-3
Formula: $C_6H_3Cl_4N$ MW: 229

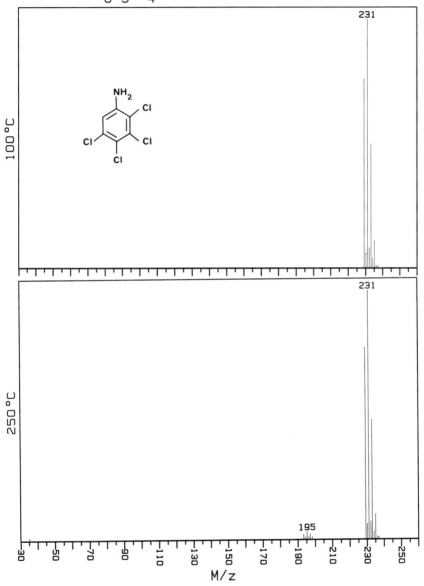

2, 3, 5, 6-Tetrachloroaniline
CAS No: 3481-20-7
Formula: $C_6H_3Cl_4N$ MW: 229

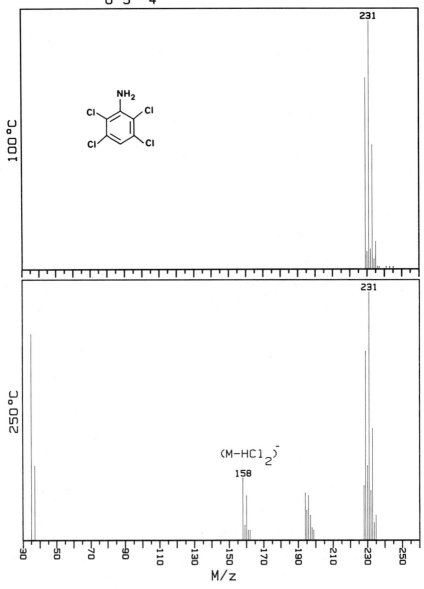

Pentachloroaniline
CAS No: 527-20-8
Formula: $C_6H_2Cl_5N$ MW: 263

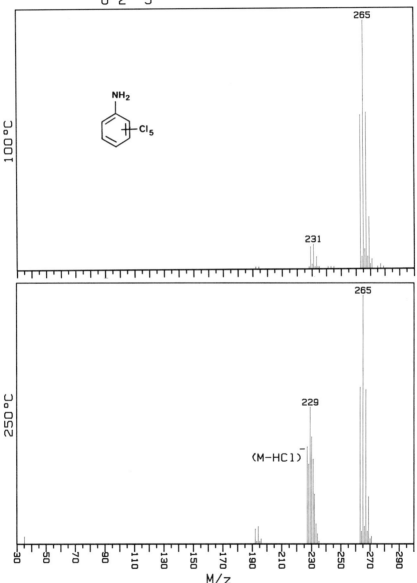

Pentafluoroaniline
CAS No: 771-60-8
Formula: $C_6H_2F_5N$ MW: 183

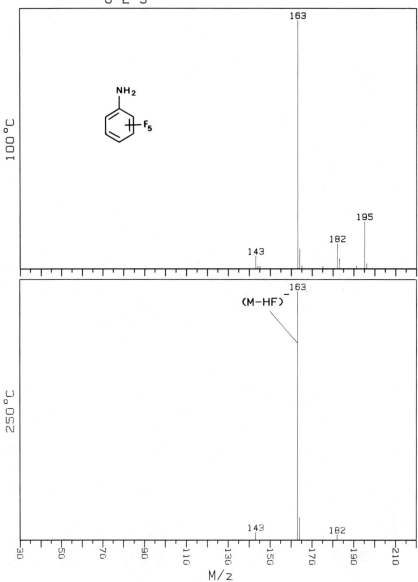

a, a-Dichlorotoluene
CAS No: 29797-40-8
Formula: $C_7H_6Cl_2$ MW: 160

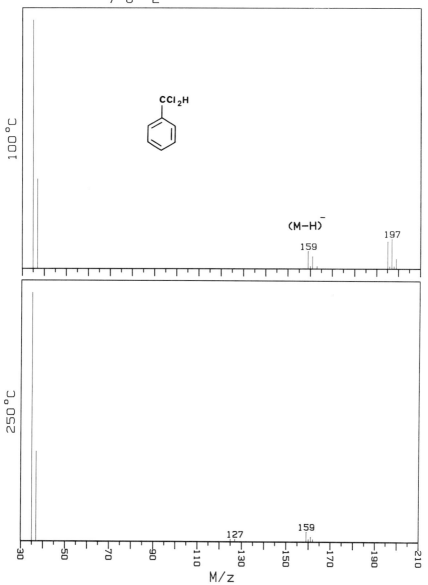

100 °C

CCl$_2$H

(M–H)⁻

159

197

250 °C

127

159

30   50   70   90   110   130   150   170   190   210

M/z

a, 4-Dichlorotoluene
CAS No: 104-83-6
Formula: $C_7H_6Cl_2$ MW: 160

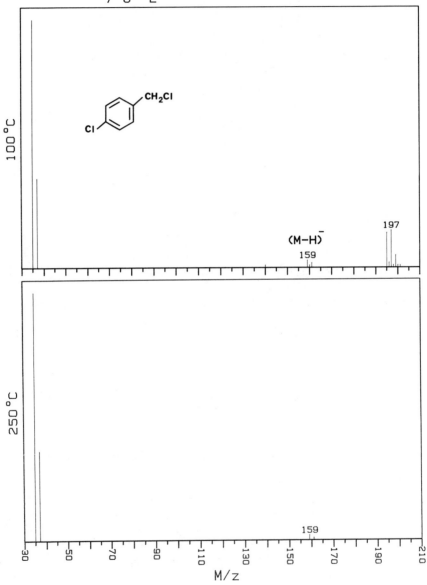

a, 2-Dichlorotoluene
CAS No: 611-19-8
Formula: C₇H₆Cl₂   MW: 160

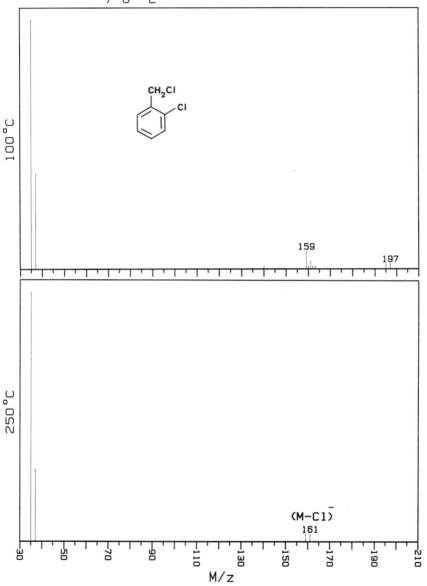

a, a, a-Trichlorotoluene
CAS No: 98-07-7
Formula: $C_7H_5Cl_3$ MW: 194

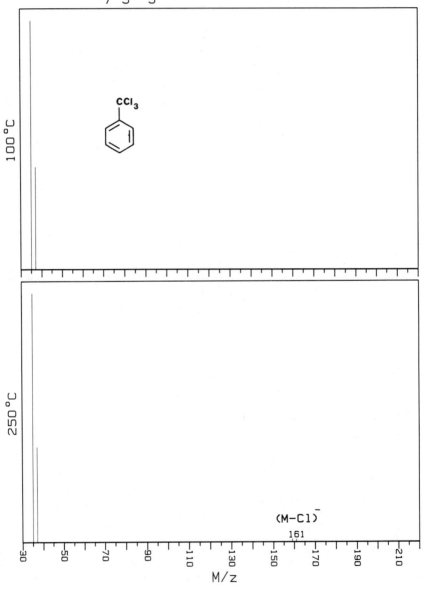

2, 4, 5-Trichlorotoluene
CAS No: 6639-30-1
Formula: $C_7H_5Cl_3$ MW: 194

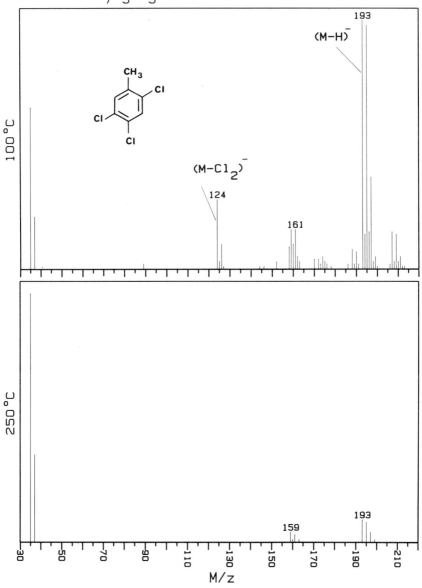

a, 2, 6-Trichlorotoluene
CAS No: 2014-83-7
Formula: $C_7H_5Cl_3$ MW: 194

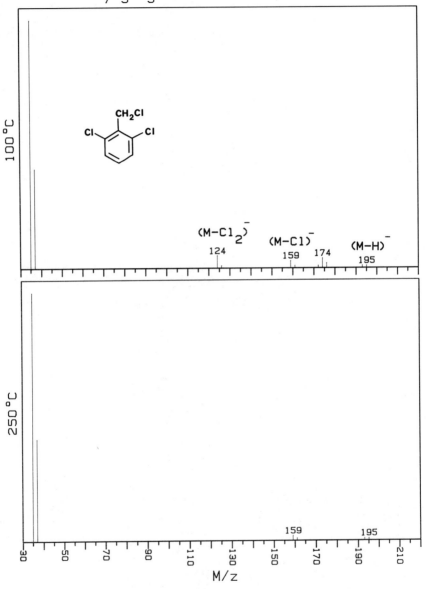

α, α, 2, 6-Tetrachlorotoluene
CAS No: 81-19-6
Formula: $C_7H_4Cl_4$ MW: 228

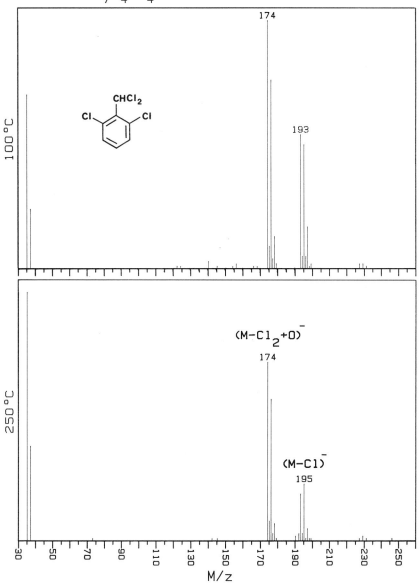

Pentafluorotoluene
CAS No: 771-56-2
Formula: C$_7$H$_3$F$_5$ MW: 182

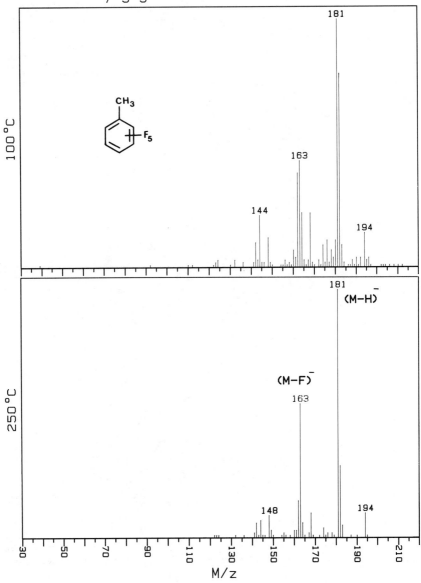

Pentachlorotoluene
CAS No: 877-11-2
Formula: $C_7H_3Cl_5$ MW: 262

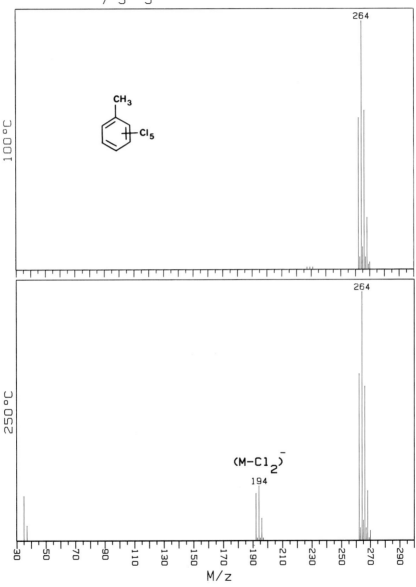

Pentabromotoluene
CAS No: 87-83-2
Formula: $C_7H_3Br_5$ MW: 482

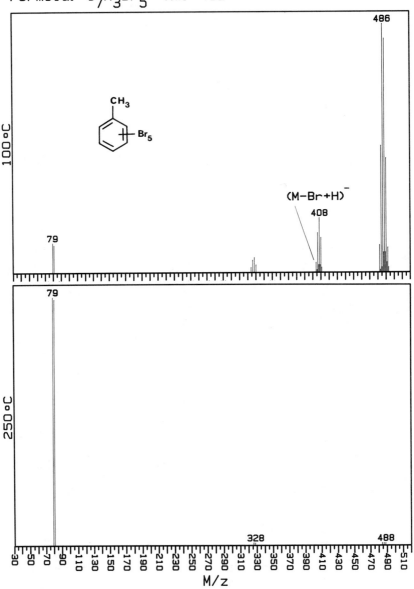

2, 4, 5, 6-Tetrachloro-m-xylene
CAS No: 877-09-8
Formula: $C_8H_6Cl_4$ MW: 242

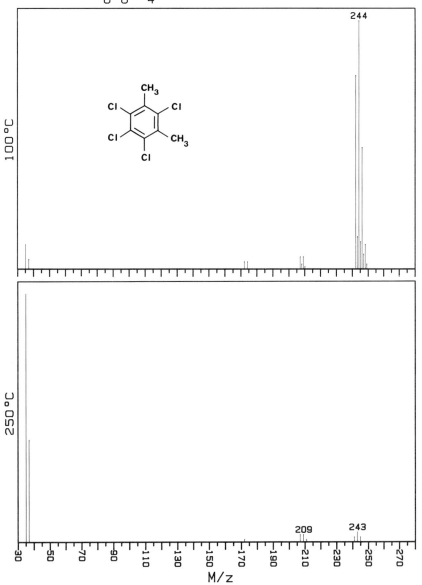

a, a′, 2, 3, 5, 6-Hexachloro-p-xylene
CAS No: 1079-17-0
Formula: $C_8H_4Cl_6$ MW: 310

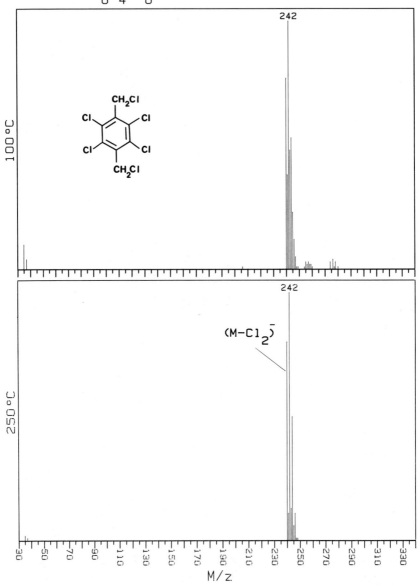

a, a, a, a', a', a'-Hexachloro-p-xylene
CAS No: 68-36-0
Formula: $C_8H_4Cl_6$ MW: 310

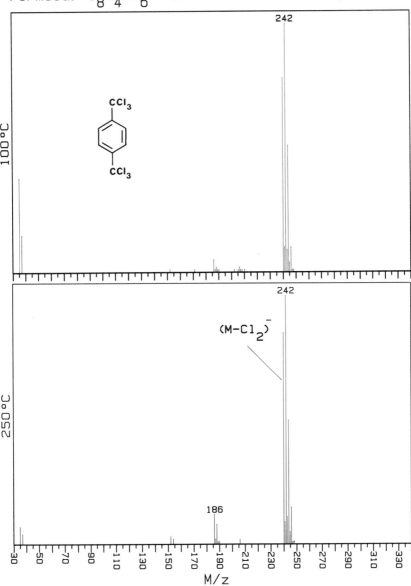

a, a, a, a', a', a'-Hexachloro-m-xylene
CAS No: 881-99-2
Formula: $C_8H_4Cl_6$ MW: 310

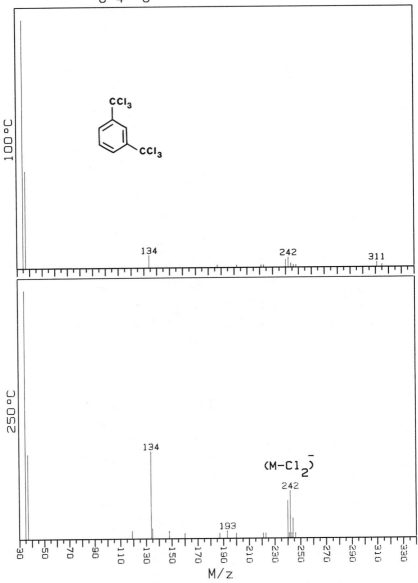

1,2-Dinitrobenzene
CAS No: 528-29-0
Formula: $C_6H_4N_2O_4$ MW: 168

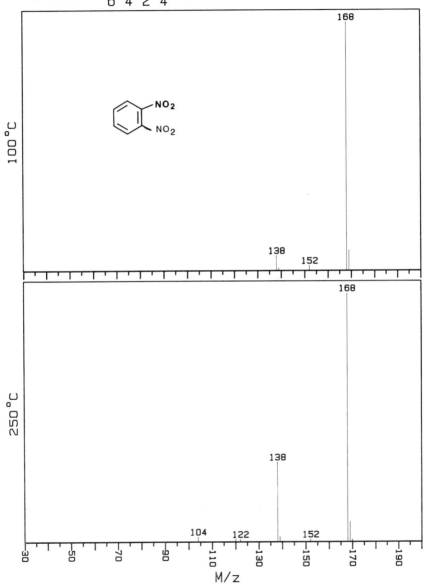

2-Nitrobenzonitrile
CAS No: 612-24-8
Formula: $C_7H_4N_2O_2$ MW: 148

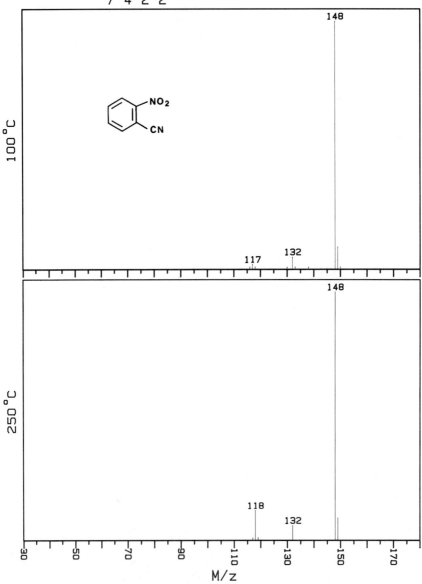

2-Nitrobenzotrifluoride
CAS No: 384-22-5
Formula: $C_7H_4F_3NO_2$ MW: 191

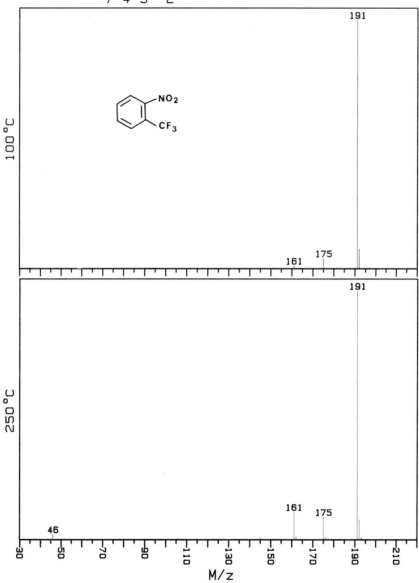

2-Nitrobenzoic acid, methyl ester
CAS No: 606-27-9
Formula: $C_8H_7NO_4$ MW: 181

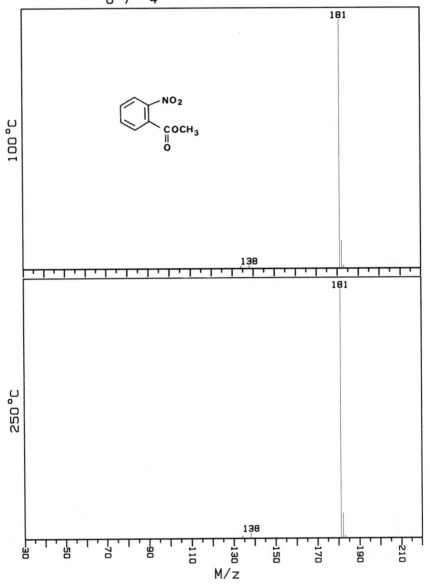

2-Nitrophenol, acetyl derivative
CAS No: 610-69-5
Formula: $C_8H_7NO_4$ MW: 181

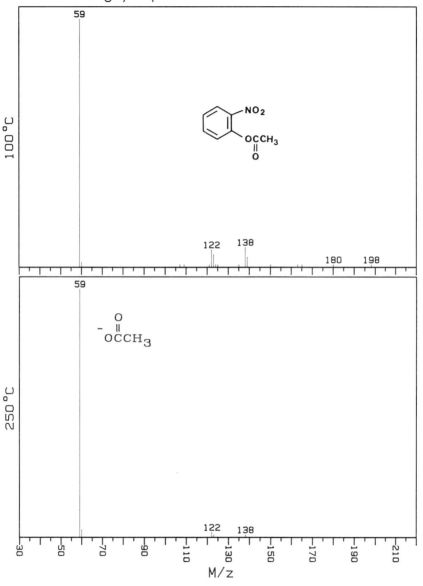

2-Nitrobenzaldehyde
CAS No: 522-89-6
Formula: $C_7H_5NO_3$ MW: 151

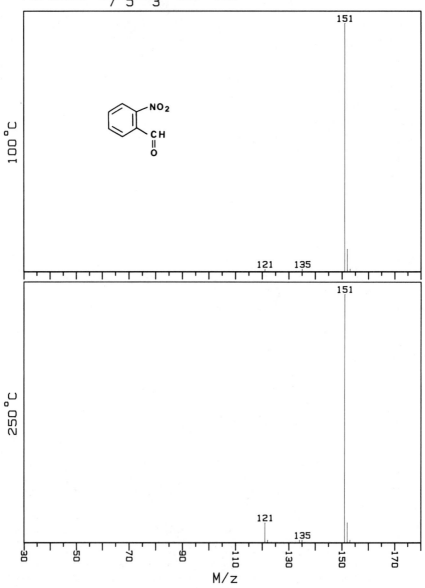

2-Nitroanisole
CAS No: 91-23-6
Formula: $C_7H_7NO_3$ MW: 153

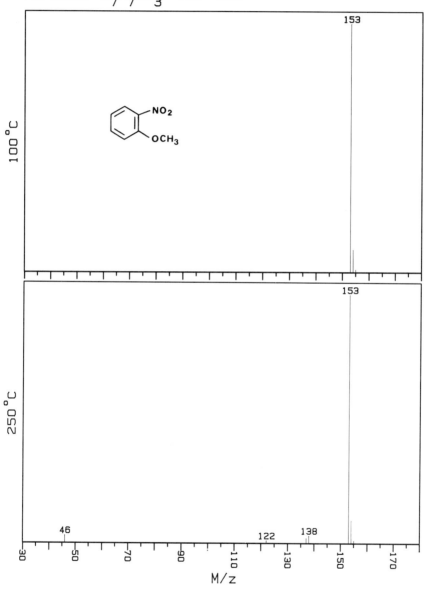

2-Nitrophenol
CAS No: 88-75-5
Formula: $C_6H_5NO_3$ MW: 139

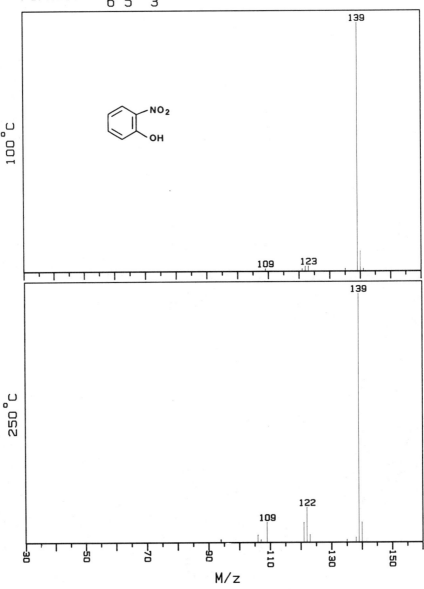

2-Nitrotoluene
CAS No: 88-72-2
Formula: $C_7H_7NO_2$ MW: 137

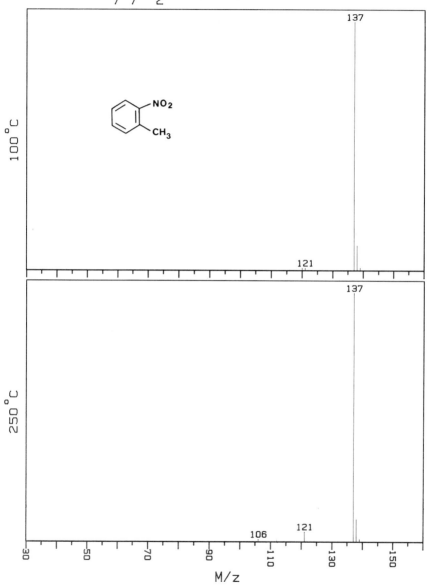

2-Nitroaniline
CAS No: 88-74-4
Formula: $C_6H_6N_2O_2$ MW: 138

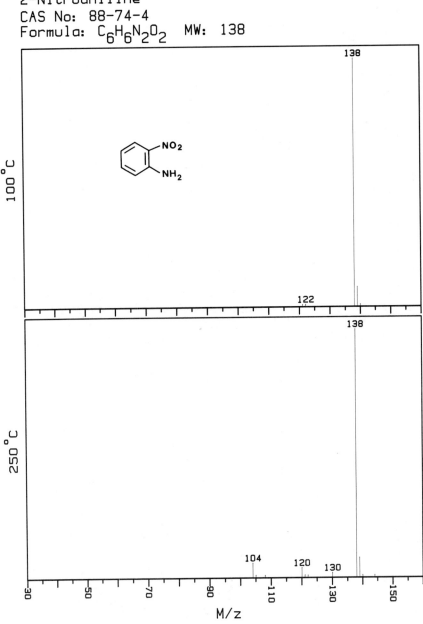

1,3-Dinitrobenzene
CAS No: 99-65-0
Formula: $C_6H_4N_2O_4$ MW: 168

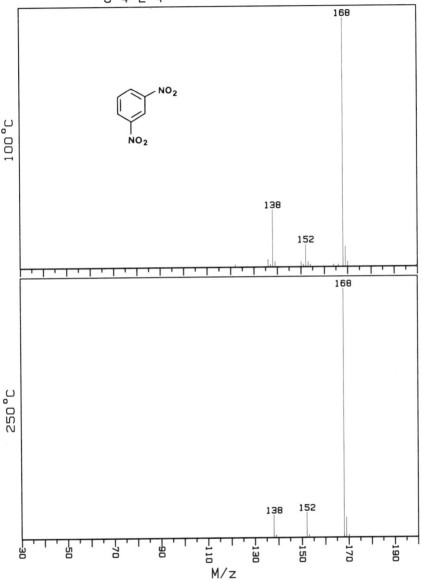

3-Nitrobenzonitrile
CAS No: 619-24-9
Formula: $C_7H_4N_2O_2$ MW: 148

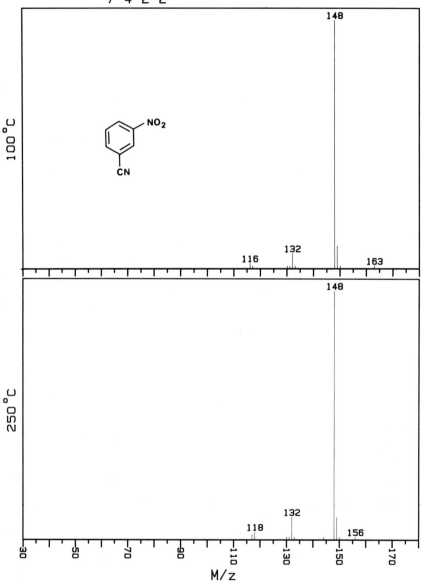

3-Nitrobenzotrifluoride
CAS No: 98-46-4
Formula: $C_7H_4F_3NO_2$ MW: 191

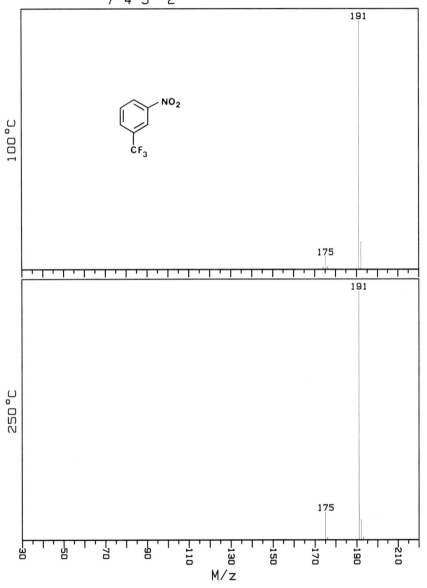

3-Nitrobenzoic acid, methyl ester
CAS No: 618-95-1
Formula: $C_8H_7NO_4$ MW: 181

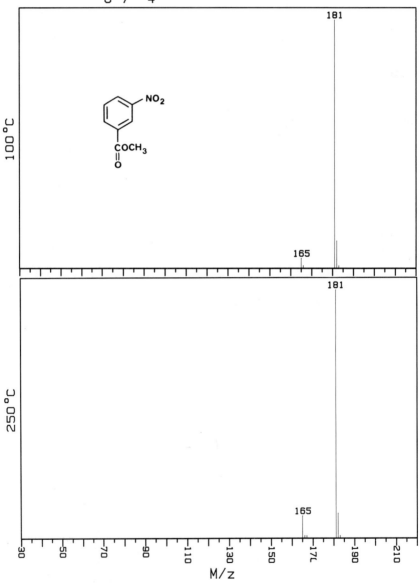

3-Nitrophenol, acetyl derivative
CAS No: 1523-06-4
Formula: $C_8H_7NO_4$ MW: 181

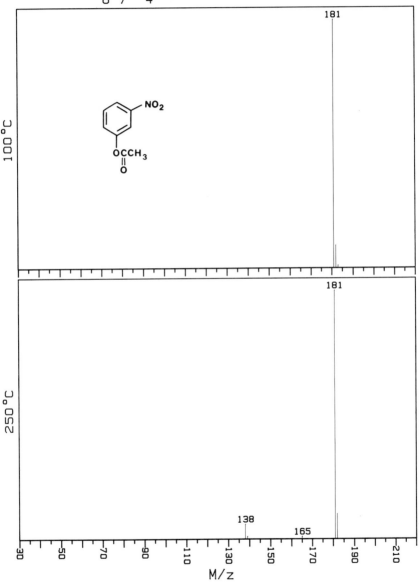

3-Nitrobenzaldehyde
CAS No: 99-61-6
Formula: $C_7H_5NO_3$ MW: 151

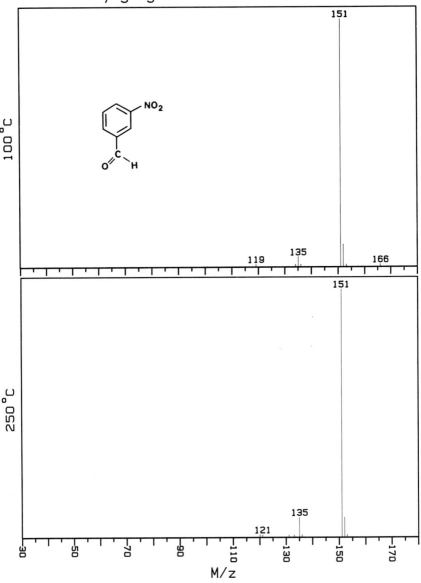

3-Nitrophenol
CAS No: 554-84-7
Formula: $C_6H_5NO_3$ MW: 139

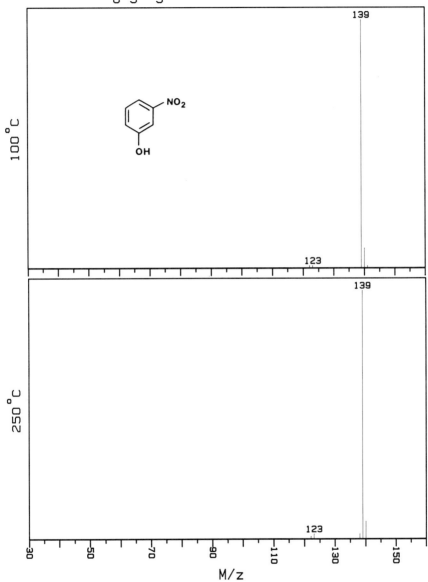

3-Nitrotoluene
CAS No: 99-08-1
Formula: $C_7H_7NO_2$ MW: 137

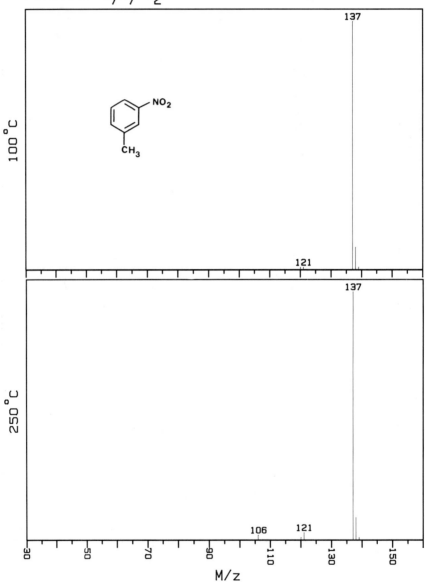

3-Nitroaniline
CAS No: 99-09-2
Formula: $C_6H_6N_2O_2$ MW: 138

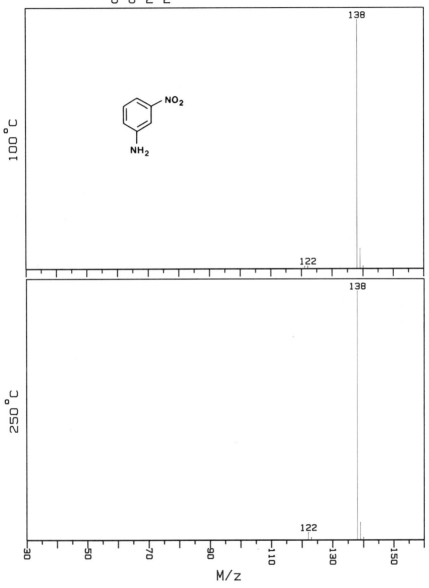

1,4-Dinitrobenzene
CAS No: 100-25-4
Formula: $C_6H_4N_2O_4$ MW: 168

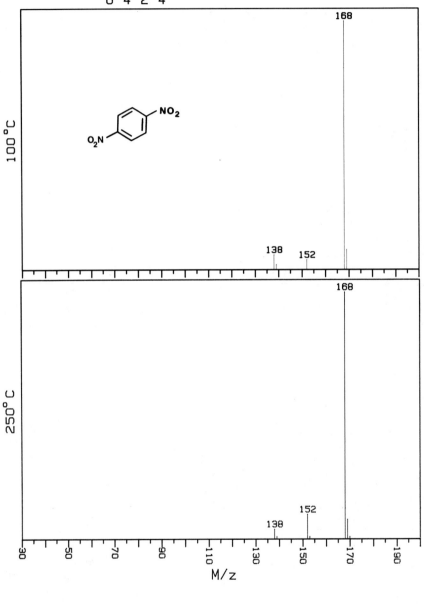

4-Nitrobenzonitrile
CAS No: 619-72-7
Formula: $C_7H_4N_2O_2$ MW: 148

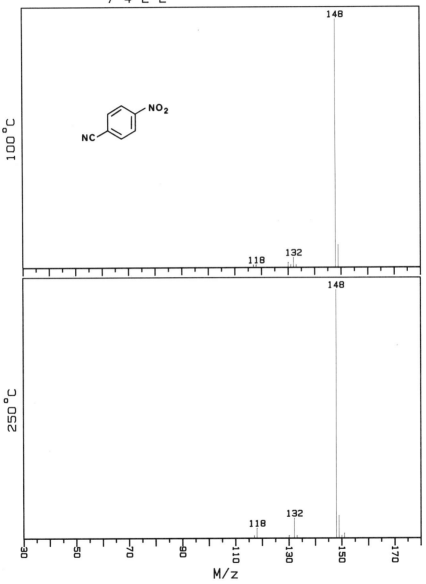

4-Nitrobenzotrifluoride
CAS No: 402-54-0
Formula: $C_7H_4F_3NO_2$ MW: 191

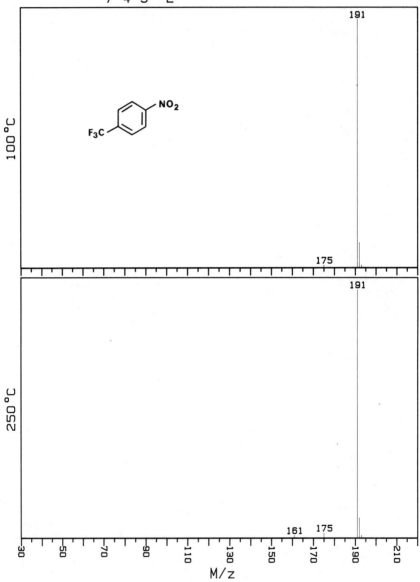

4-Nitrobenzoic acid, methyl ester
CAS No: 619-50-1
Formula: $C_8H_7NO_4$ MW: 181

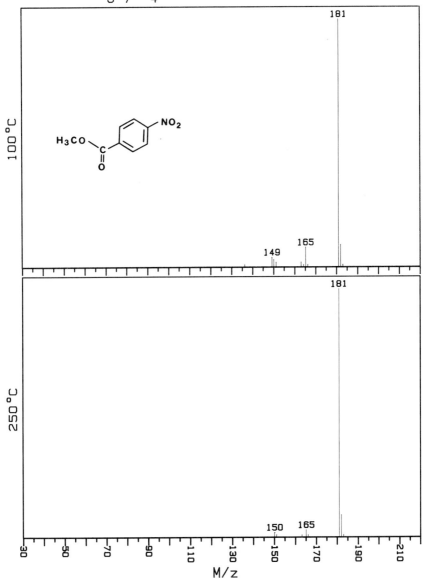

4-Nitrophenol, acetyl derivative
CAS No: 830-03-5
Formula: $C_8H_7NO_4$ MW: 181

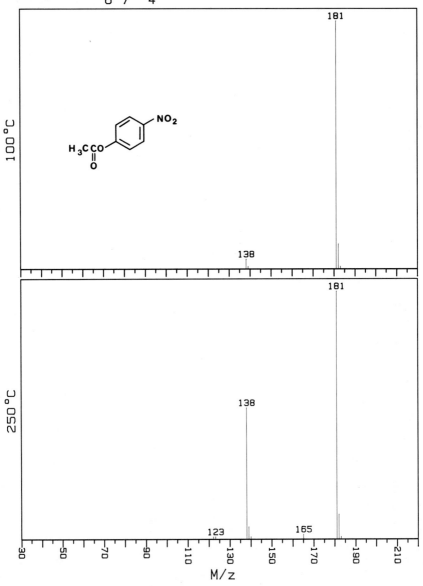

4-Nitrobenzaldehyde
CAS No: 555-16-8
Formula: $C_7H_5NO_3$ MW: 151

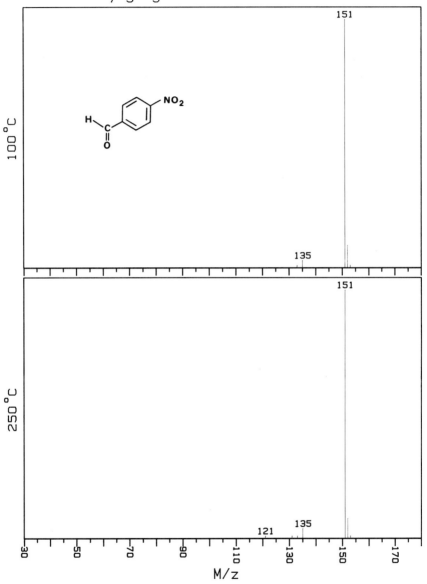

4-Nitroanisole
CAS No: 100-17-4
Formula: $C_7H_7NO_3$ MW: 153

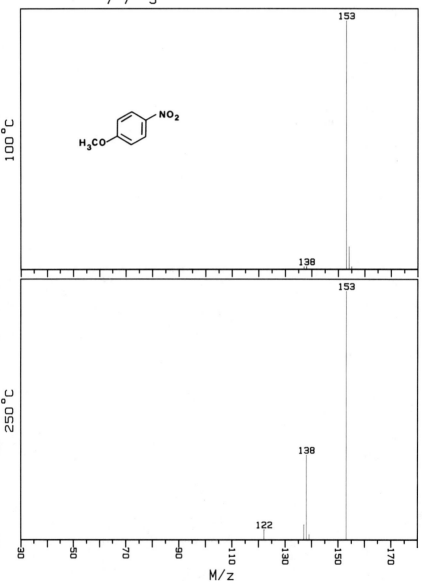

4-Nitrophenol
CAS No: 100-02-7
Formula: $C_6H_5NO_3$ MW: 139

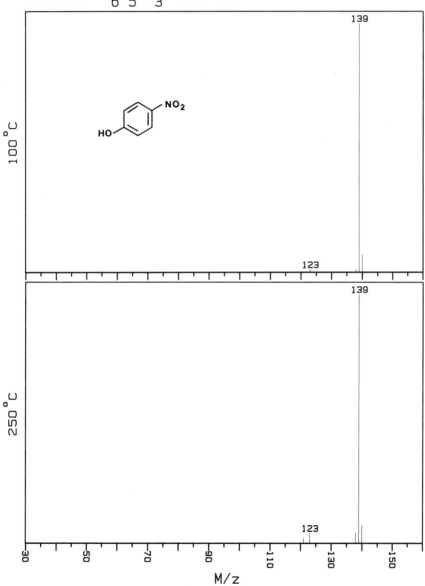

4-Nitrotoluene
CAS No: 99-99-0
Formula: $C_7H_7NO_2$ MW: 137

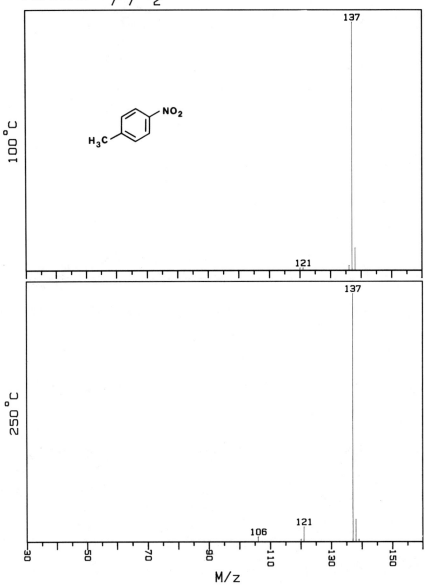

4-Nitroaniline
CAS No: 100-01-6
Formula: $C_6H_6N_2O_2$ MW: 138

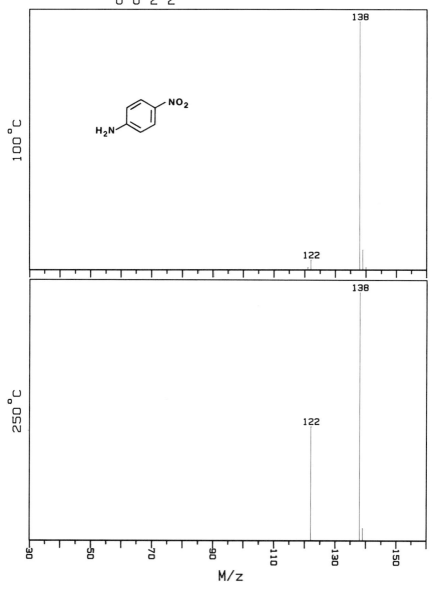

4-Methyl-2-nitrophenol
CAS No: 119-33-5
Formula: $C_7H_7NO_3$ MW: 153

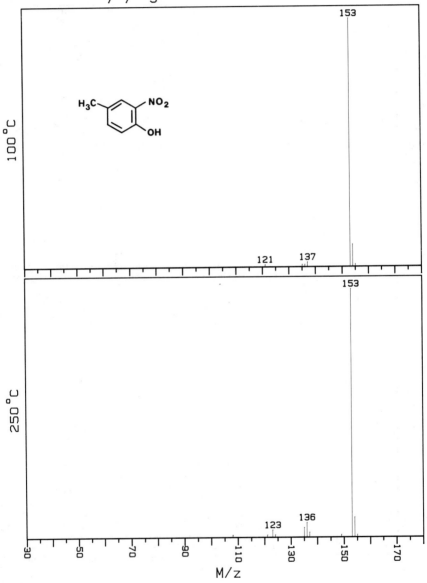

3,5-Dinitrobenzonitrile
CAS No: 4110-35-4
Formula: $C_7H_3N_3O_4$ MW: 193

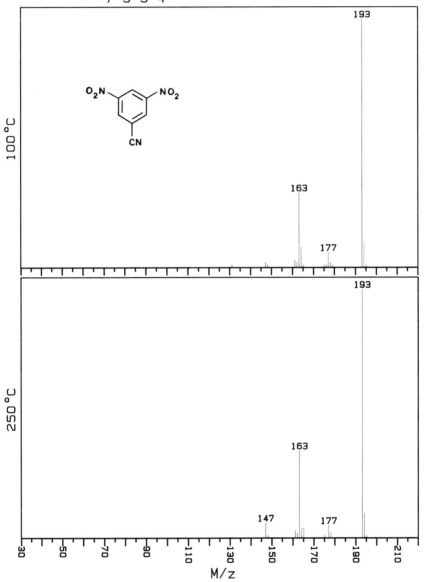

3,5-Dinitrobenzotrifluoride
CAS No: 401-99-0
Formula: $C_7H_3F_3N_2O_4$ MW: 236

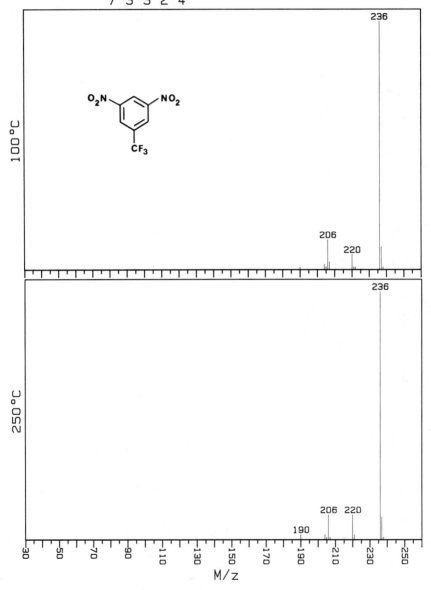

3,5-Dinitrobenzamide
CAS No: 121-81-3
Formula: $C_7H_5N_3O_5$ MW: 211

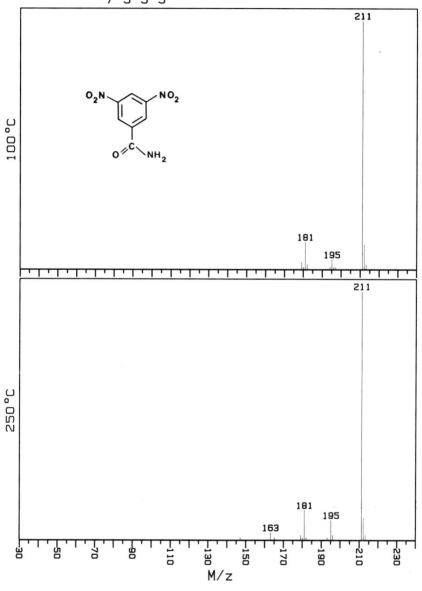

3,5-Dinitrobenzoic acid, methyl ester
CAS No: 2702-58-1
Formula: $C_8H_6N_2O_6$, MW: 226

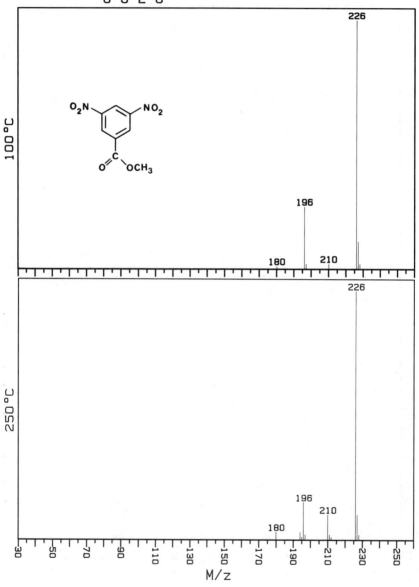

3,5-Dinitrobenzyl alcohol
CAS No: 71022-43-0
Formula: $C_7H_6N_2O_5$ MW: 198

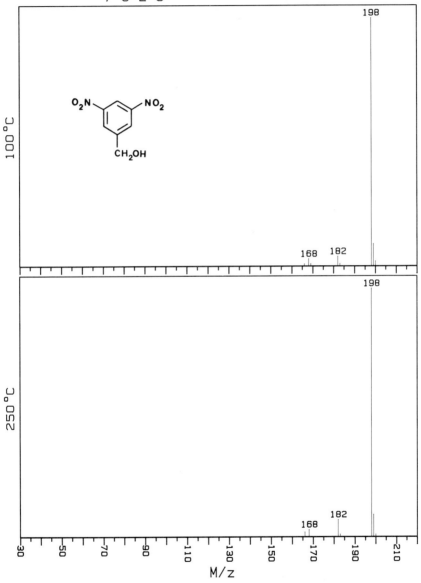

3,5-Dinitroaniline
CAS No: 618-87-1
Formula: $C_6H_5N_3O_4$ MW: 183

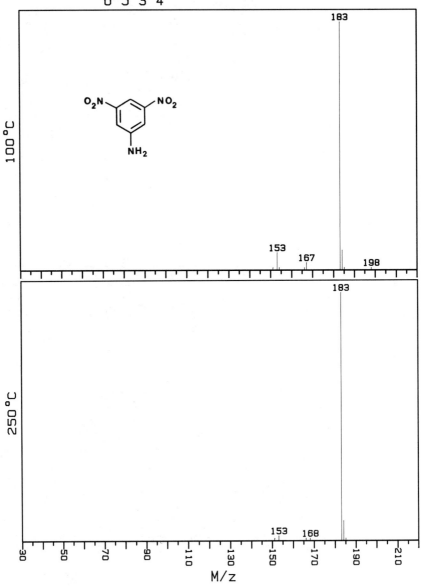

4-Chloro-3,5-dinitrobenzonitrile
CAS No: 1930-72-9
Formula: $C_7H_2ClN_3O_4$ MW: 227

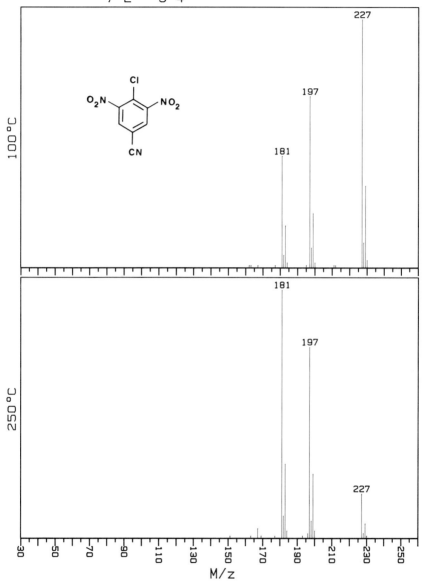

4-Chloro-3,5-dinitrobenzotrifluoride
CAS No: 393-75-9
Formula: $C_7H_2ClF_3N_2O_4$ MW: 270

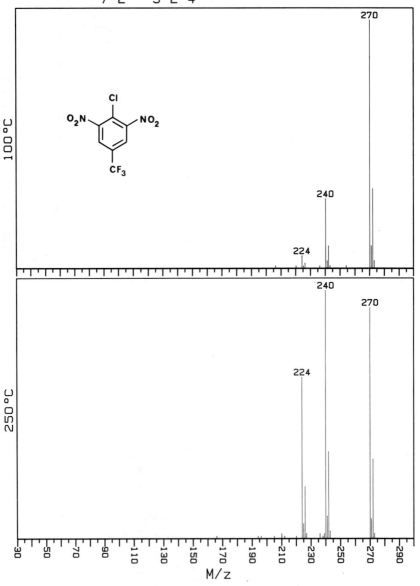

4-Chloro-3,5-dinitrobenzoic acid, methyl ester
CAS No: 118-97-8
Formula: $C_8H_5ClN_2O_6$ MW: 260

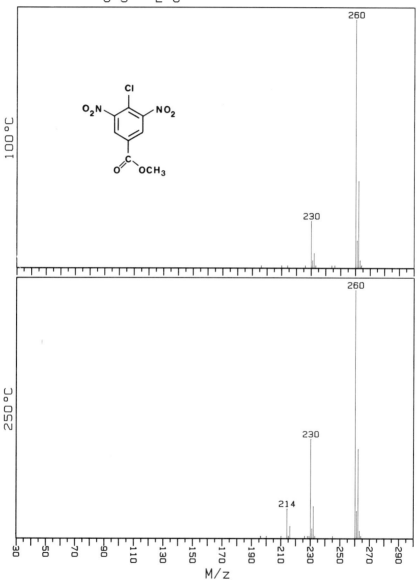

1-Chloro-2, 4-dinitrobenzene
CAS No: 97-00-7
Formula: $C_6H_3ClN_2O_4$ MW: 202

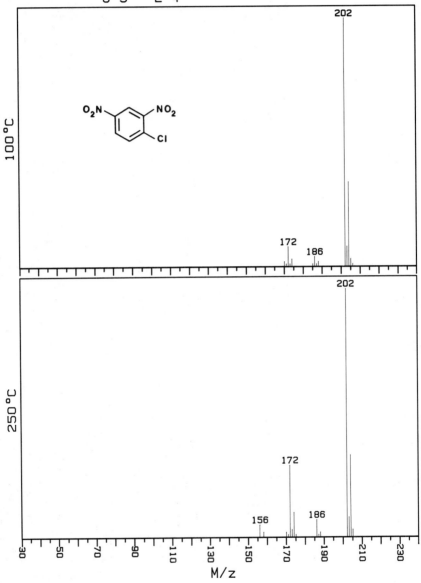

2,4-Dinitrophenol, acetyl derivative
CAS No: 4232-27-3
Formula: $C_8H_6N_2O_6$ MW: 226

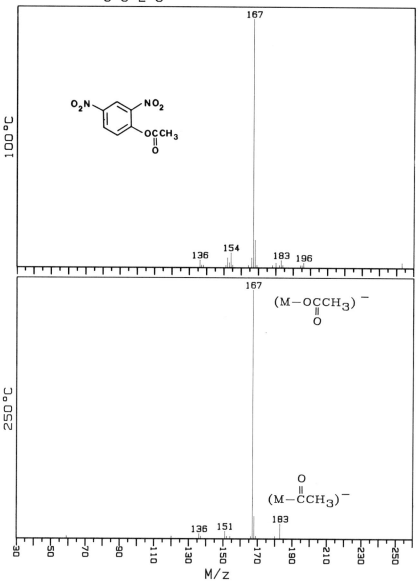

2,4-Dinitroanisole
CAS No: 119-27-7
Formula: $C_7H_6N_2O_5$, MW: 198

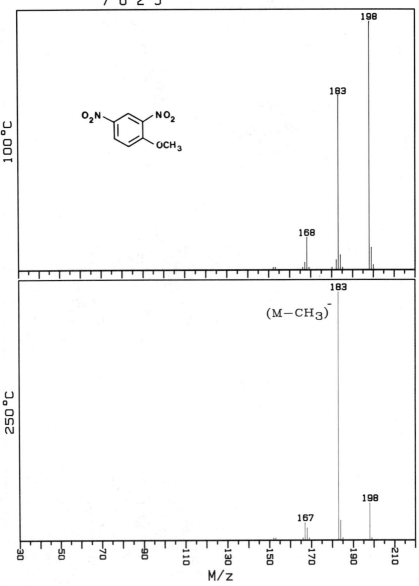

2,4-Dinitrophenol
CAS No: 51-28-5
Formula: $C_6H_4N_2O_5$ MW: 184

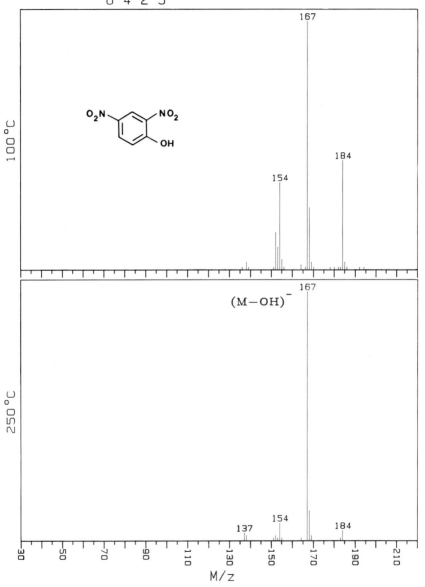

2, 4-Dinitrotoluene
CAS No: 121-14-2
Formula: $C_7H_6N_2O_4$ MW: 182

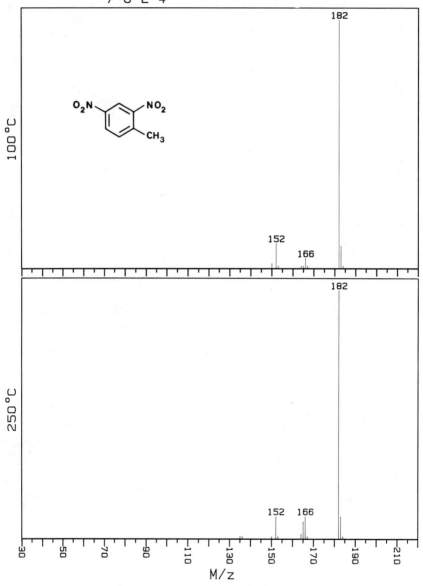

N-(2-Chloroethyl)-2,4-dinitroaniline
CAS No: not avail.
Formula: $C_8H_8ClN_3O_4$ MW: 245

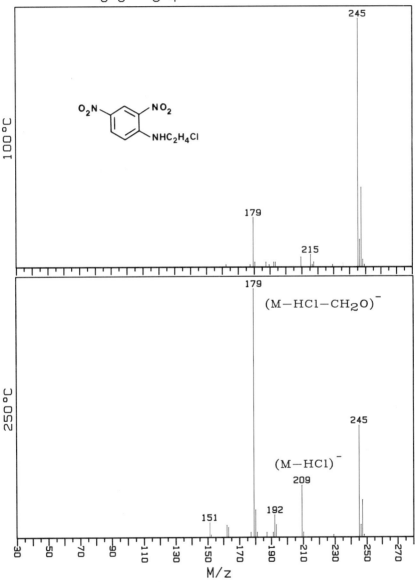

N, N-Diethyl-2, 4-dinitroaniline
CAS No: 837-64-9
Formula: $C_{10}H_{13}N_3O_4$, MW: 239

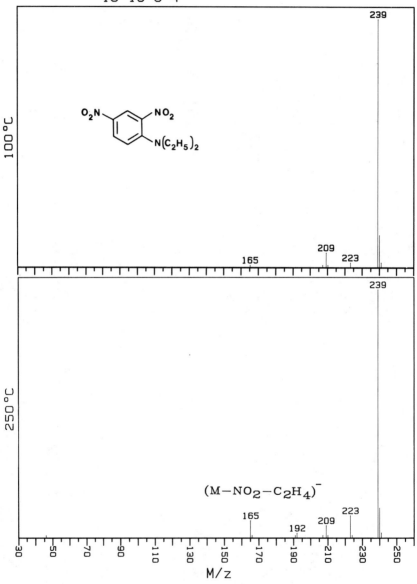

2,4-Dinitro-N,N-dipropylaniline
CAS No: 54718-72-8
Formula: $C_{12}H_{17}N_3O_4$ MW: 267

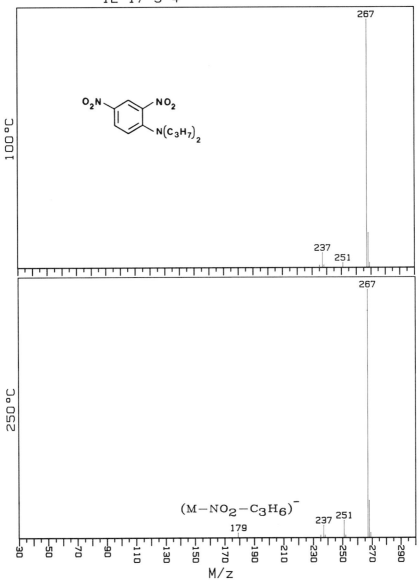

2,4-Dinitro-N-methylaniline
CAS No: 2044-88-4
Formula: $C_7H_7N_3O_4$ MW: 197

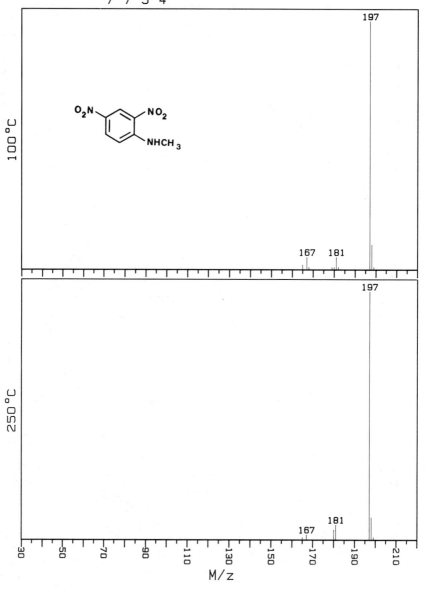

2,4-Dinitroaniline
CAS No: 97-02-9
Formula: C₆H₅N₃O₄   MW: 183

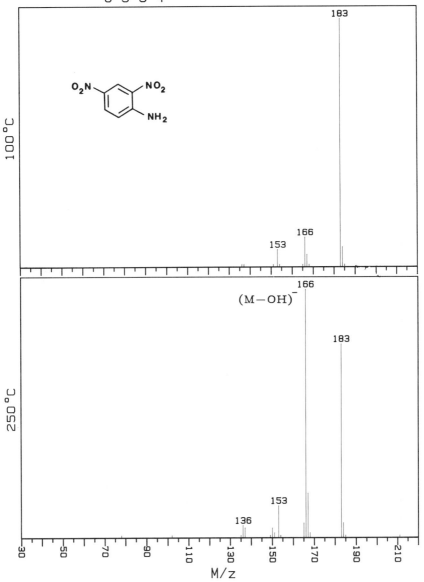

4,5-Dichloro-2-nitroaniline
CAS No: 6641-64-1
Formula: $C_6H_4Cl_2N_2O_2$ MW: 206

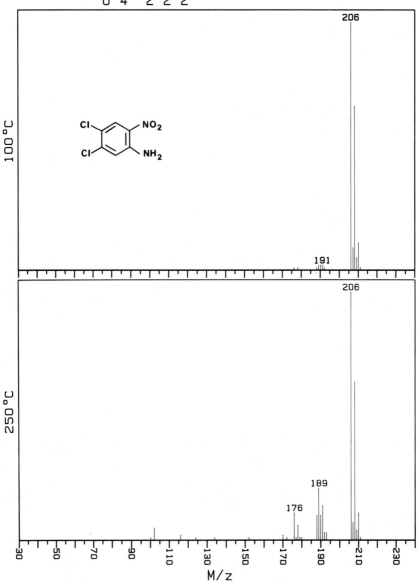

2, 4-Dichloro-6-nitroaniline
CAS No: 2683-43-4
Formula: $C_6H_4Cl_2N_2O_2$ MW: 206

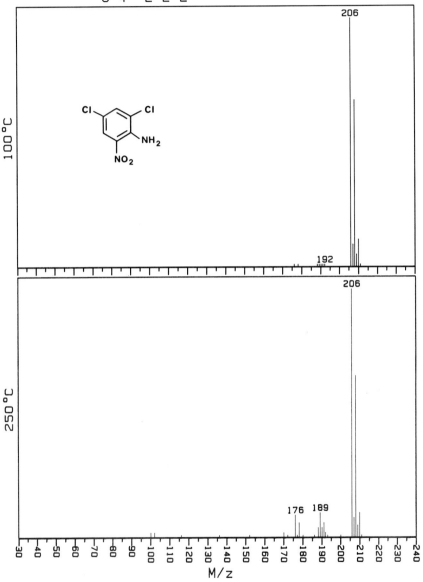

2,6-Dichloro-4-nitroaniline
CAS No: 99-30-9
Formula: $C_6H_4Cl_2N_2O_2$ MW: 206

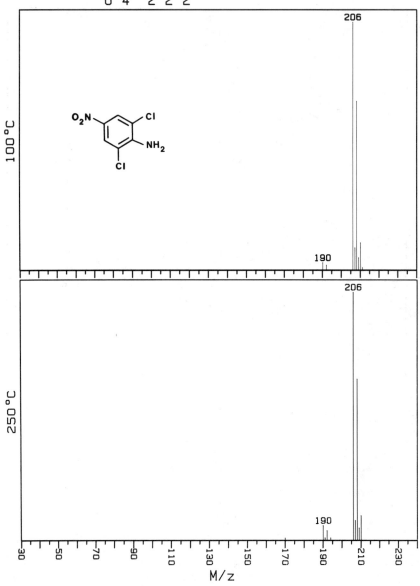

6-Chloro-2,4-dinitroaniline
CAS No: 3531-19-9
Formula: $C_6H_4ClN_3O_4$ MW: 217

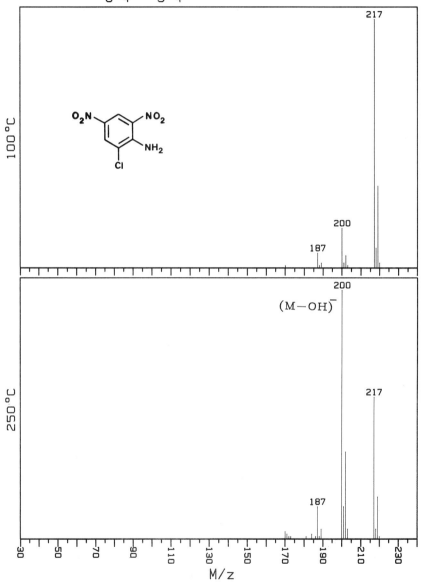

6-Bromo-2,4-dinitroaniline
CAS No: 1817-73-8
Formula: $C_6H_4BrN_3O_4$ MW: 261

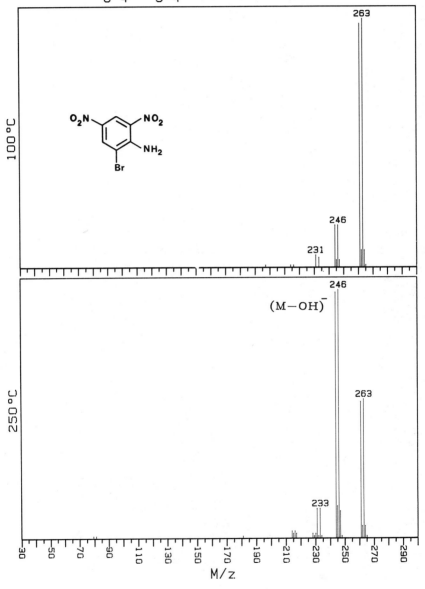

2-Amino-4,6-dinitrotoluene
CAS No: 35572-78-2
Formula: $C_7H_7N_3O_4$ MW: 197

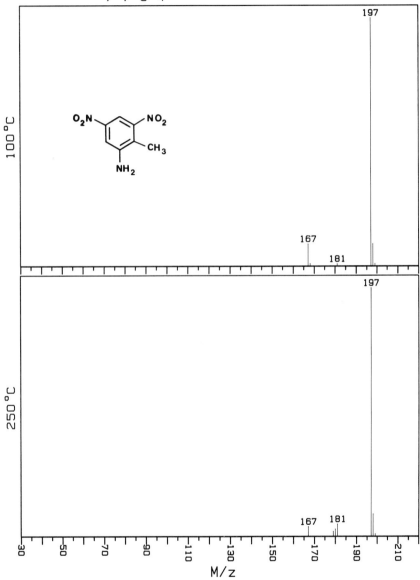

2-Chloro-5-nitrobenzotrifluoride
CAS No: 777-37-7
Formula: $C_7H_3ClF_3NO_2$ MW: 225

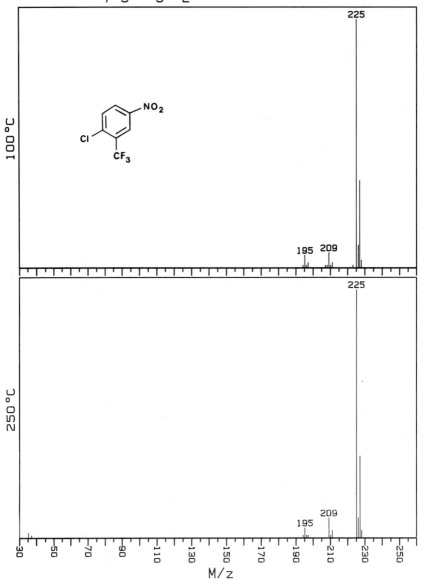

4-Chloro-3-nitrobenzotrifluoride
CAS No: 121-17-5
Formula: $C_7H_3ClF_3NO_2$ MW: 225

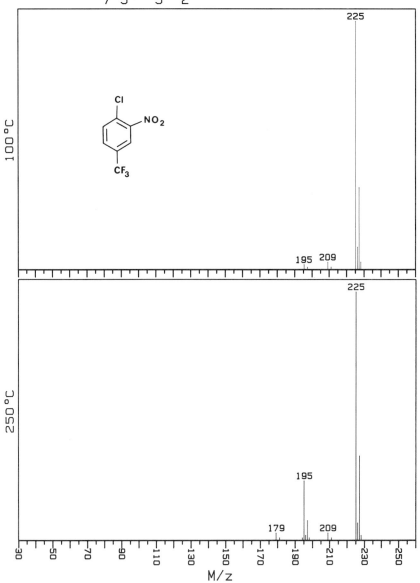

5-Chloro-2-nitrobenzotrifluoride
CAS No: 118-83-2
Formula: $C_7H_3ClF_3NO_2$ MW: 225

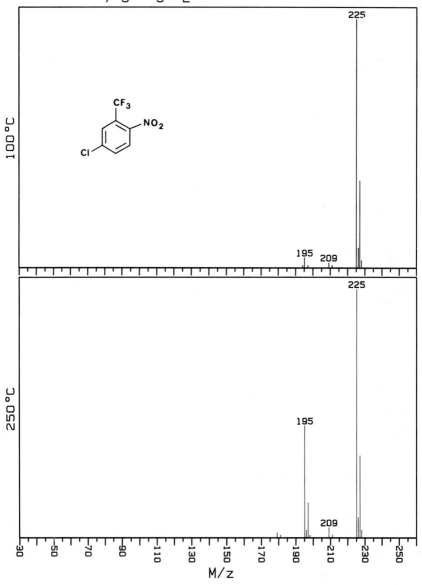

2,6-Dinitro-p-creosol
CAS No: 609-93-8
Formula: $C_7H_6N_2O_5$ MW: 198

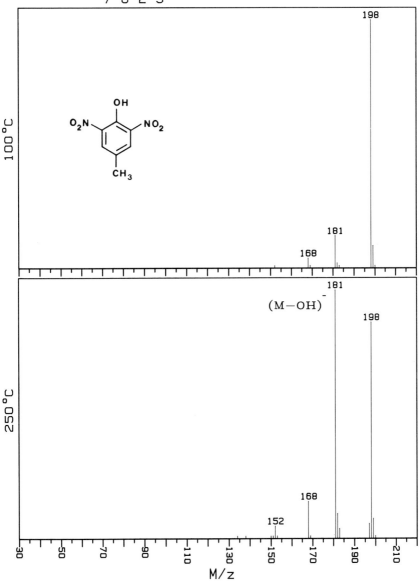

2,6-Dinitro-4-(trifluoromethyl)phenol
CAS No: 393-77-1
Formula: $C_7H_3F_3N_2O_5$ MW: 252

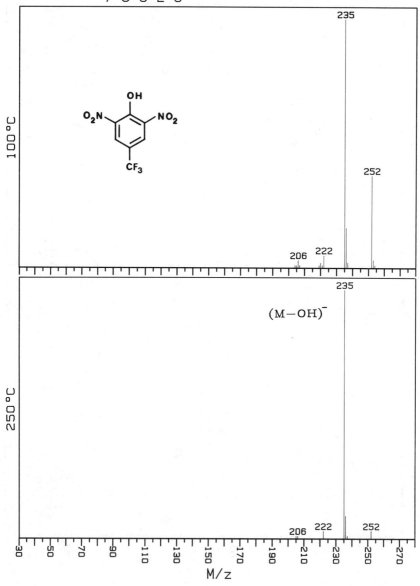

2,6-Dinitrotoluene
CAS No: 606-20-2
Formula: $C_7H_6N_2O_4$, MW: 182

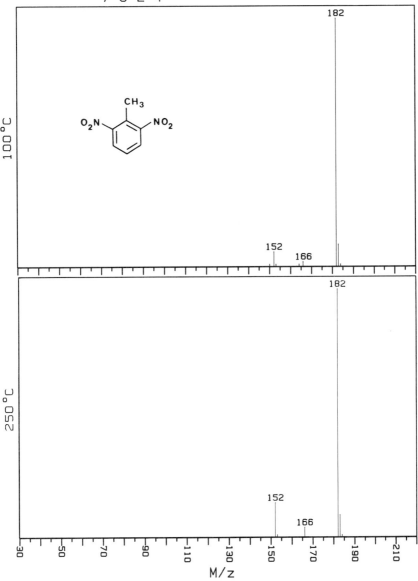

2,6-Dinitro-N-ethylaniline
CAS No: 7449-13-0
Formula: $C_8H_9N_3O_4$ MW: 211

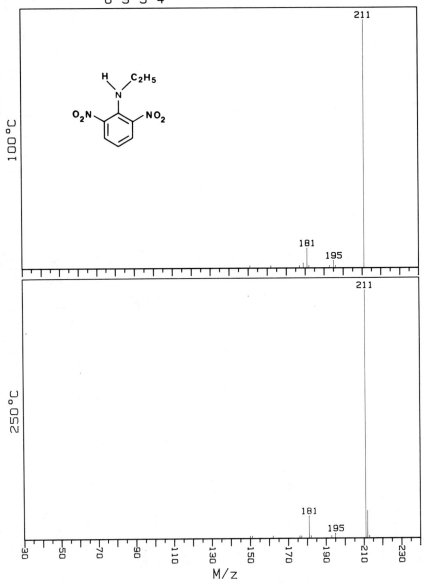

2,6-Dinitroaniline
CAS No: 606-22-4
Formula: $C_6H_5N_3O_4$ MW: 183

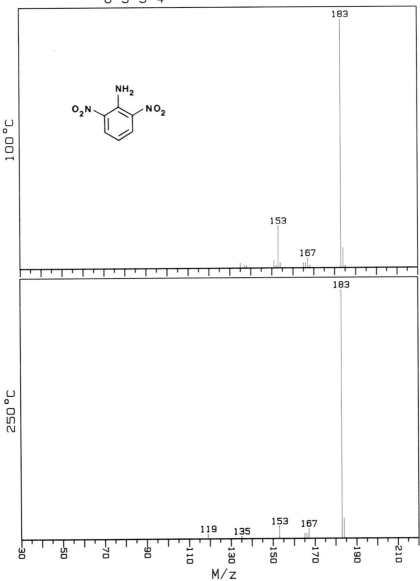

2,3-Dinitrotoluene
CAS No: 602-01-7
Formula: $C_7H_6N_2O_4$, MW: 182

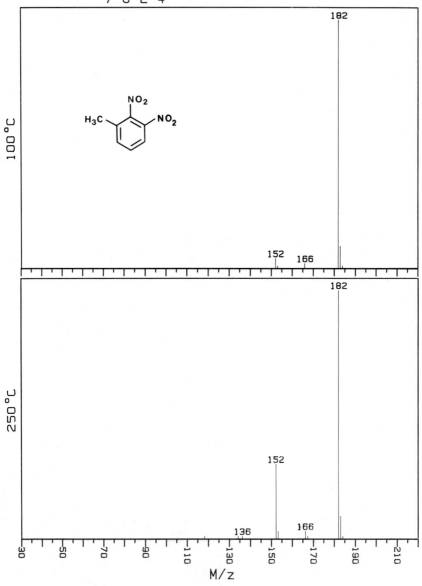

2,4-Dichloro-3,5-dinitrobenzonitrile

CAS No: 1930-71-8

Formula: $C_7HCl_2N_3O_4$ MW: 261

2,4-Dichloro-3,5-dinitrobenzotrifluoride
CAS No: 29091-09-6
Formula: $C_7HCl_2F_3N_2O_4$ MW: 304

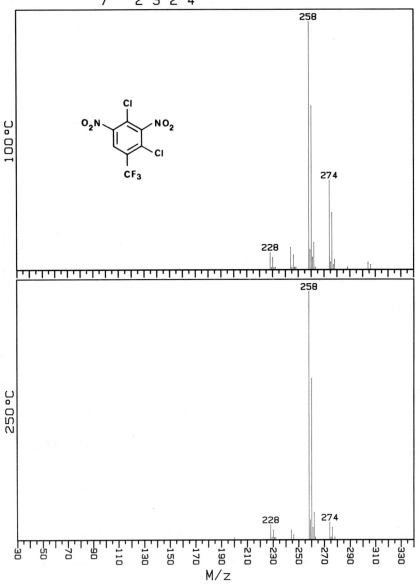

2,4-Dichloro-3,5-dinitrobenzamide
CAS No: 13550-88-4
Formula: $C_7H_3Cl_2N_3O_5$ MW: 279

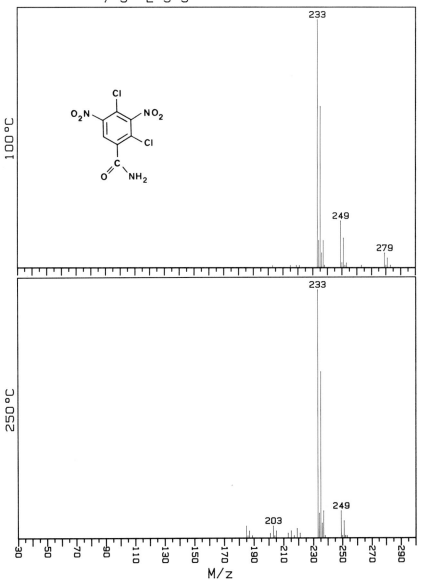

1,3-Dinitro-2,4,5-trichlorobenzene
CAS No: 2678-21-9
Formula: $C_6HCl_3N_2O_4$ MW: 270

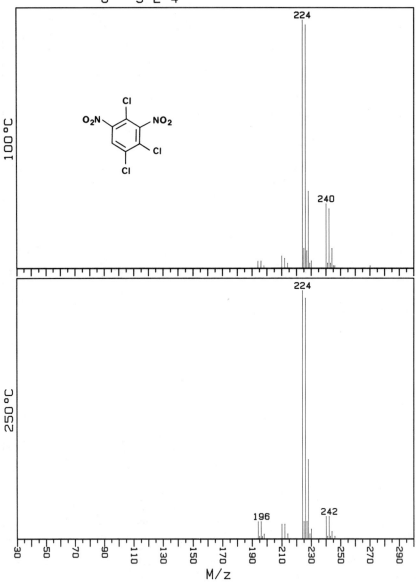

Trifluralin
CAS No: 1582-09-8
Formula: $C_{13}H_{16}F_3N_3O_4$ MW: 335

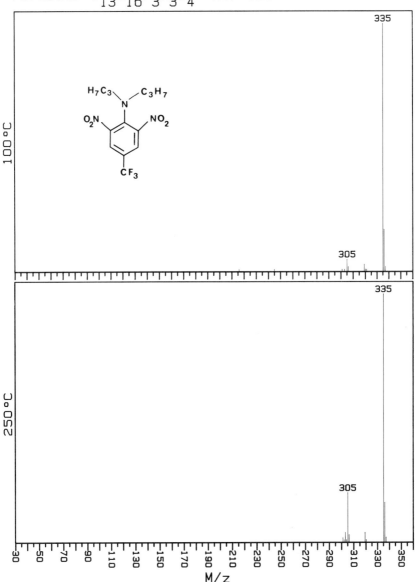

Benefin
CAS No: 1861-40-1
Formula: $C_{13}H_{16}F_3N_3O_4$ MW: 335

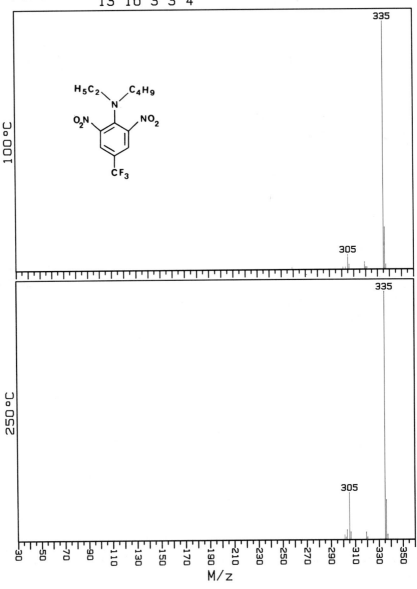

Profluralin
CAS No: 26399-36-0
Formula: $C_{14}H_{16}F_3N_3O_4$ MW: 347

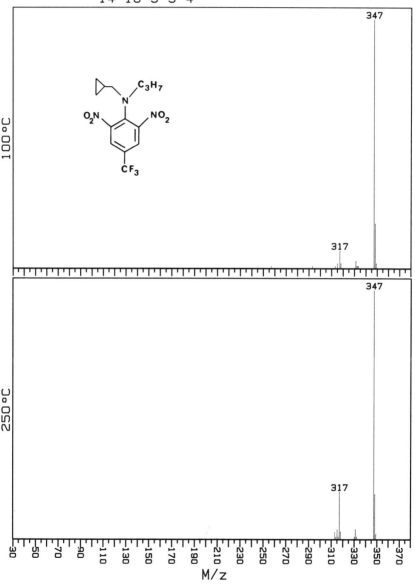

N, N-Diethyl-2, 6-dinitro-4-(trifluoromethyl)aniline
CAS No: 10223-72-0
Formula: $C_{11}H_{12}F_3N_3O_4$ MW: 307

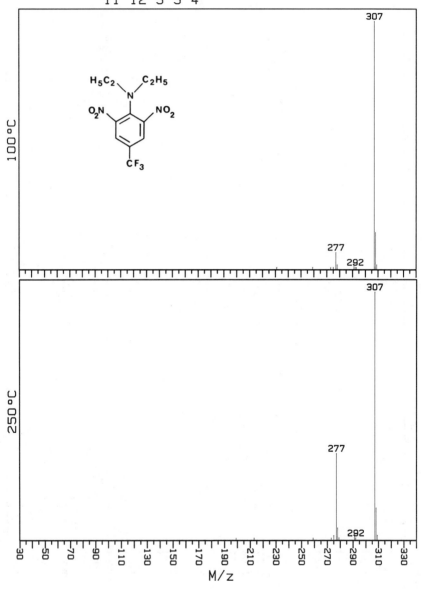

N,N-Dimethyl-2,6-dinitro-4-(trifluromethyl)aniline
CAS No: 10156-75-9
Formula: $C_9H_8F_3N_3O_4$ MW: 279

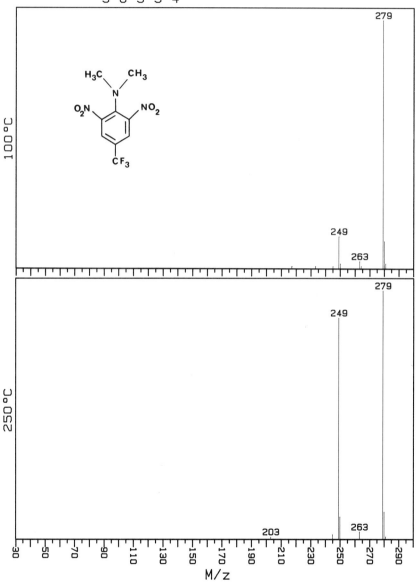

Isopropalin
CAS No: 33820-53-0
Formula: $C_{15}H_{23}N_3O_4$ MW: 309

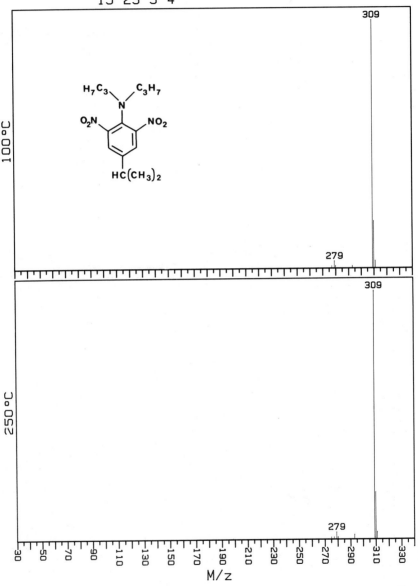

Oxyzalin
CAS No: 19044-88-3
Formula: $C_{12}H_{18}N_4O_6S$ MW: 346

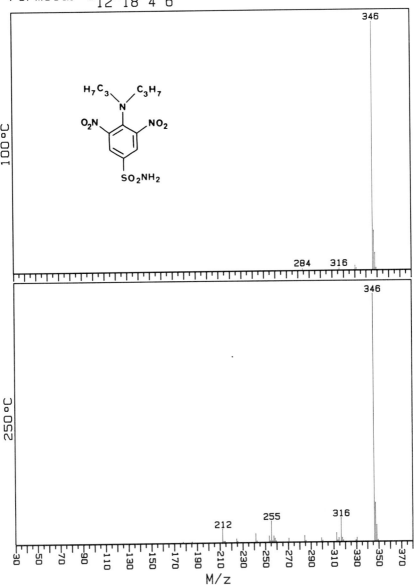

Dimethyl Oxyzalin
CAS No: 19044-94-1
Formula: $C_{14}H_{22}N_4O_6S$ MW: 374

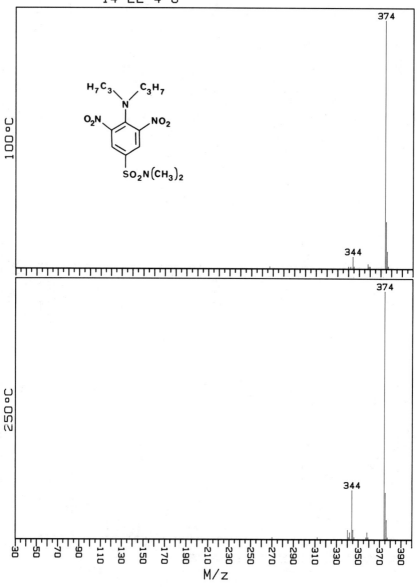

Nitralin
CAS No: 4726-14-1
Formula: $C_{13}H_{19}N_3O_6S$ MW: 345

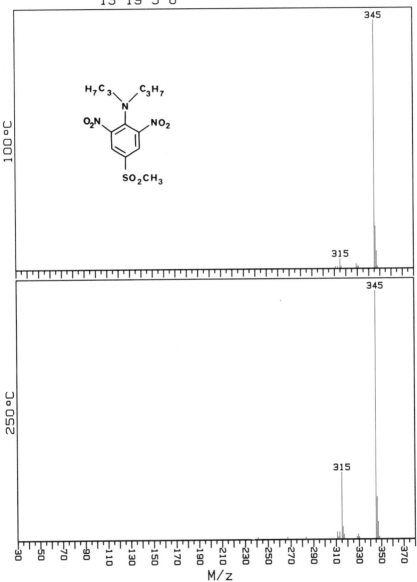

2,6-Dinitro-N,N-dipropyl-4-methoxycarbonylaniline
CAS No: 2078-11-7
Formula: $C_{14}H_{19}N_3O_6$ MW: 325

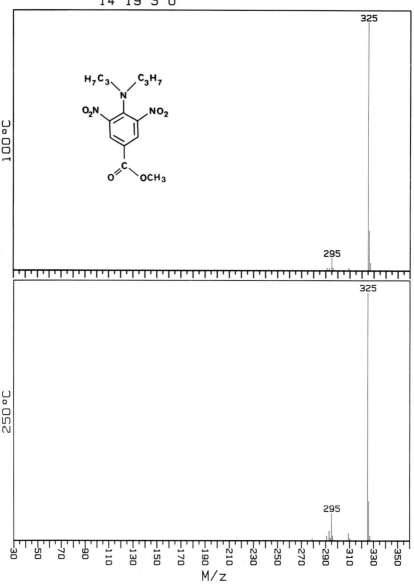

M/z

4-Cyano-2,6-dinitro-N,N-dipropylaniline
CAS No: 10228-56-5
Formula: $C_{13}H_{16}N_4O_4$ MW: 292

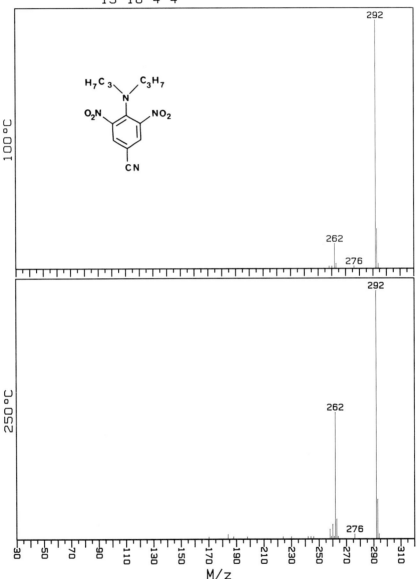

2,6-Dinitro-N-ethyl-4-(trifluoromethyl)aniline
CAS No: 10223-69-5
Formula: $C_9H_8F_3N_3O_4$ MW: 279

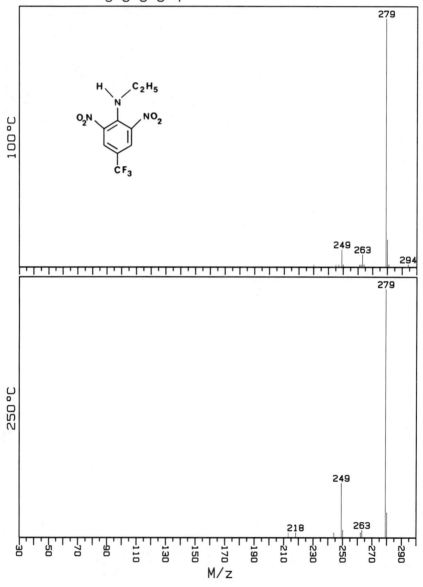

2,6-Dinitro-N-propyl-4-(trifluoromethyl)aniline
CAS No: 2077-99-8
Formula: $C_{10}H_{10}F_3N_3O_4$ MW: 293

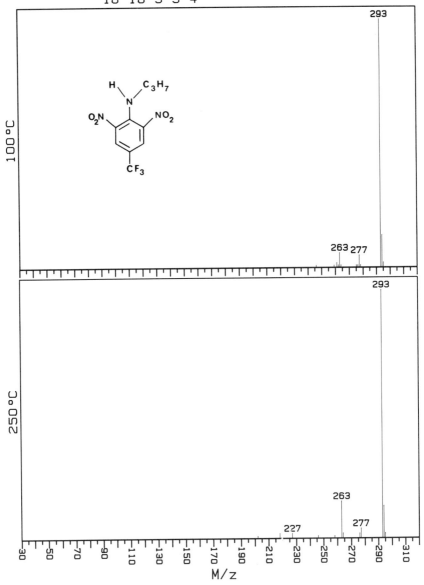

Pendimethalin
CAS No: 40318-45-4
Formula: $C_{13}H_{19}N_3O_4$ MW: 281

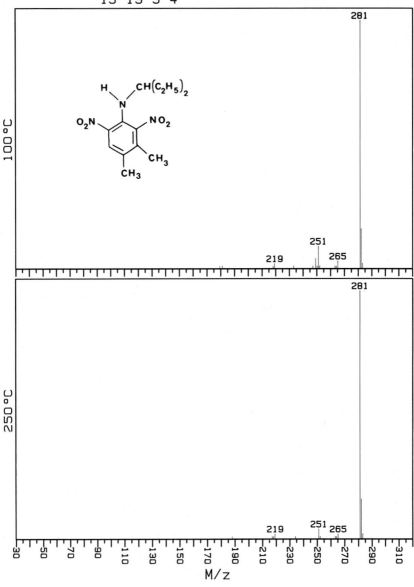

Butralin
CAS No: 33629-47-9
Formula: $C_{14}H_{21}N_3O_4$ MW: 295

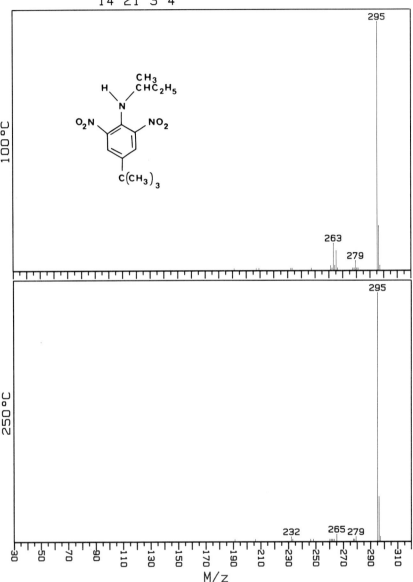

DNOC
CAS No: 534-52-1
Formula: $C_7H_6N_2O_5$ MW: 198

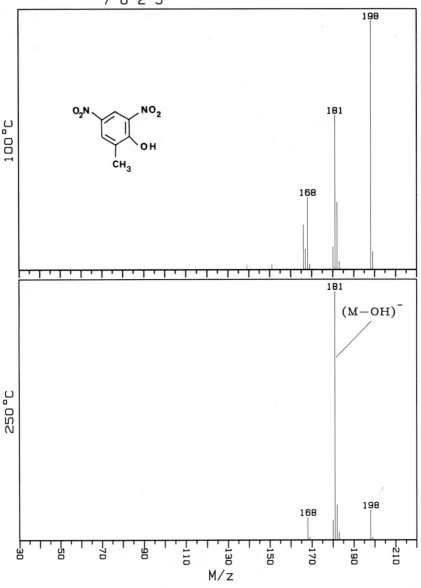

2,4-Dinitro-6-isopropylphenol
CAS No: 118-95-6
Formula: $C_9H_{10}N_2O_5$ MW: 226

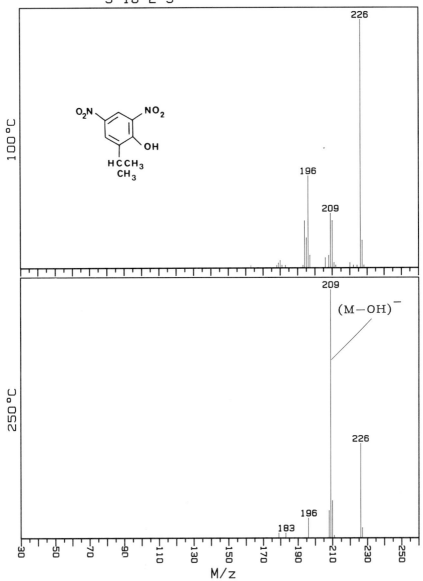

Dinoseb
CAS No: 88-85-7
Formula: $C_{10}H_{12}N_2O_5$ MW: 240

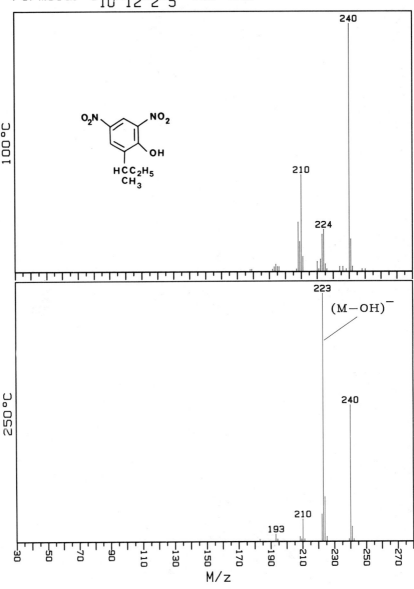

Dinoseb Acetate
CAS No: 2813-95-8
Formula: $C_{12}H_{14}N_2O_6$ MW: 282

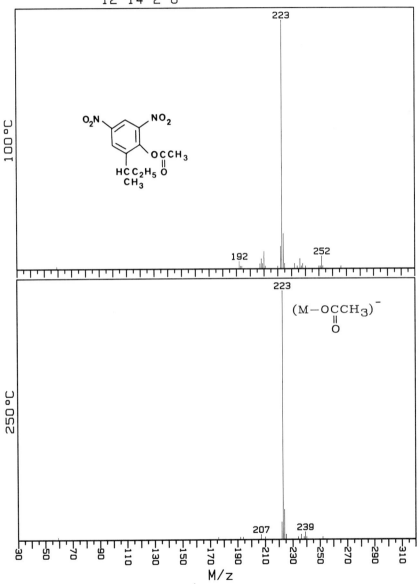

Dinocap 1
CAS No: 39300-45-3
Formula: $C_{18}H_{24}N_2O_6$ MW: 364

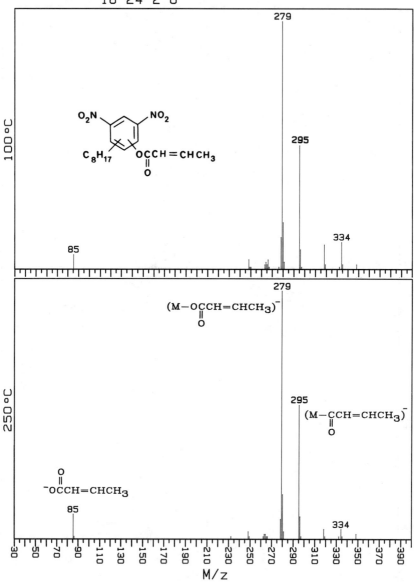

Dinocap 2
CAS No: 39300-45-3
Formula: $C_{18}H_{24}N_2O_6$ MW: 364

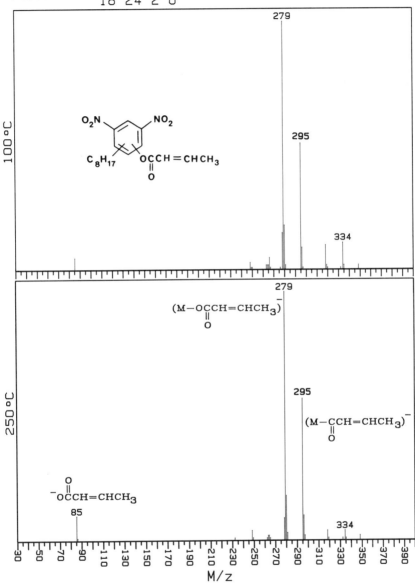

Dinocap 3
CAS No: 39300-45-3
Formula: $C_{18}H_{24}N_2O_6$ MW: 364

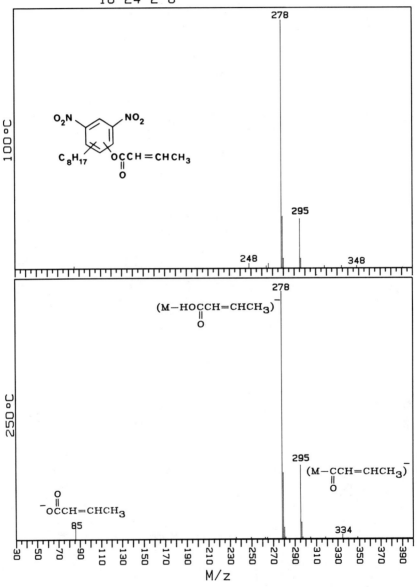

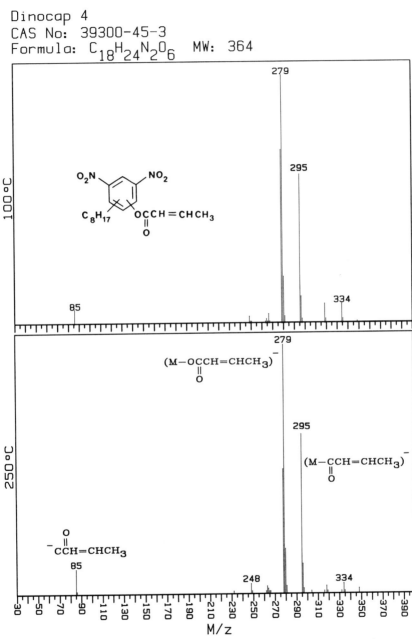

Dinocap 4
CAS No: 39300-45-3
Formula: $C_{18}H_{24}N_2O_6$ MW: 364

Dinocap 5
CAS No: 39300-45-3
Formula: $C_{18}H_{24}N_2O_6$ MW: 364

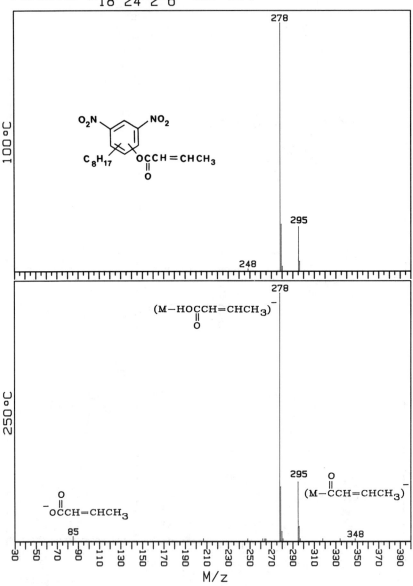

Dinitramine
CAS No: 29091-05-2
Formula: $C_{11}H_{13}F_3N_4O_4$ MW: 322

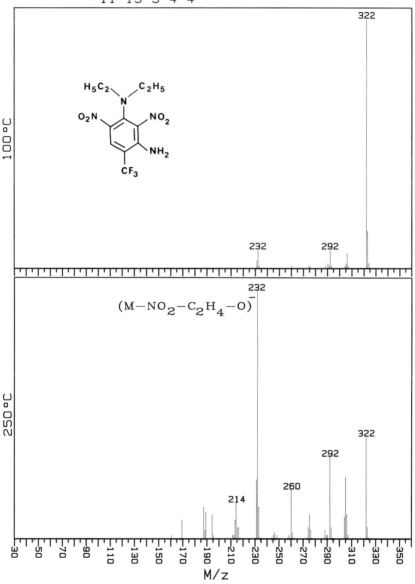

Dinitramine, acetyl derivative
CAS No: 67857-06-1
Formula: $C_{13}H_{15}F_3N_4O_5$ MW: 364

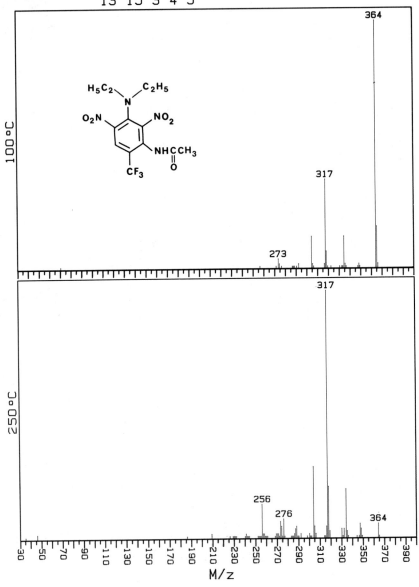

3-Chloro-N,N-dimethyl-2,6-dinitro-4-(trifluoromethyl)aniline
CAS No: 59431-66-2
Formula: $C_9H_7ClF_3N_3O_4$ MW: 313

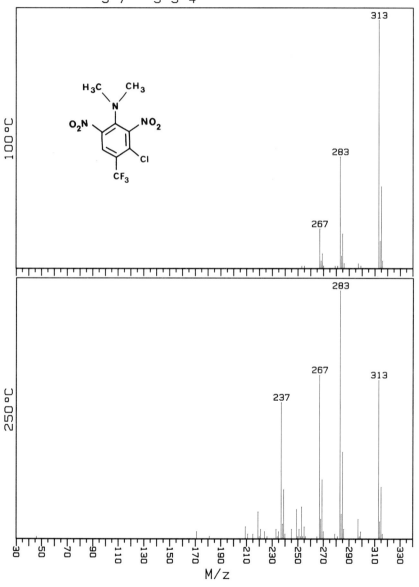

3-Chloro-N,N-diethyl-2,6-dinitro-4-(trifluoromethyl)aniline

CAS No: 36438-51-4

Formula: $C_{11}H_{11}ClF_3N_3O_4$ MW: 341

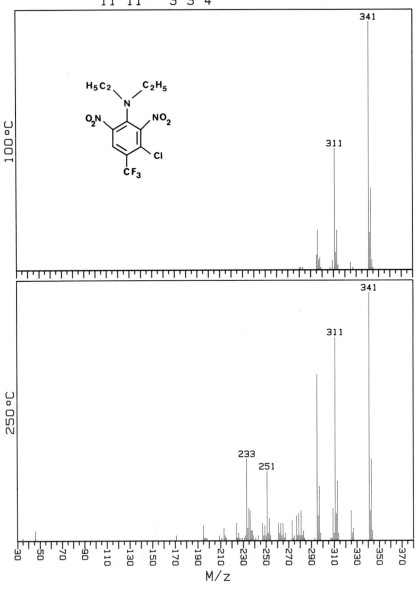

3-Chloro-2,6-dinitro-N,N-dipropyl-4-(trifluoromethyl)aniline
CAS No: not avail.
Formula: $C_{13}H_{15}ClF_3N_3O_4$ MW: 369

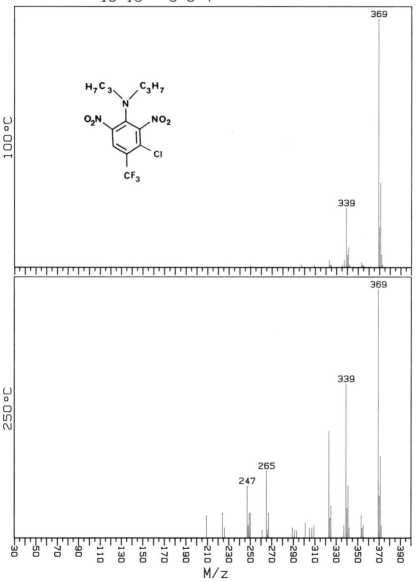

2,6-Dinitro-N,N,N,N-tetramethyl-4-(trifluoromethyl)-1,3-benzenediamine

CAS No: 29104-46-9

Formula: $C_{11}H_{13}F_3N_4O_4$ MW: 322

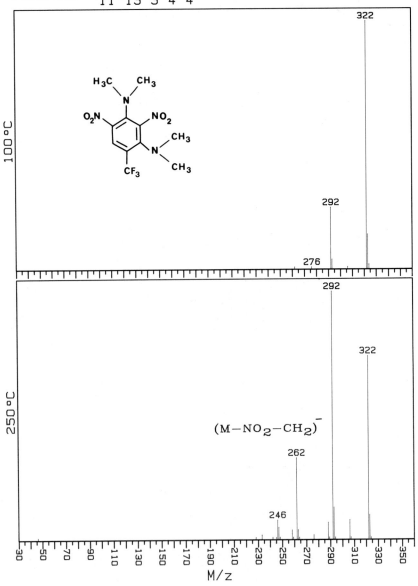

4-Cyano-2,6-dinitro-N,N,N,N-tetramethyl-1,3-benzenediamine
CAS No: not avail.
Formula: $C_{11}H_{13}N_5O_4$ MW: 279

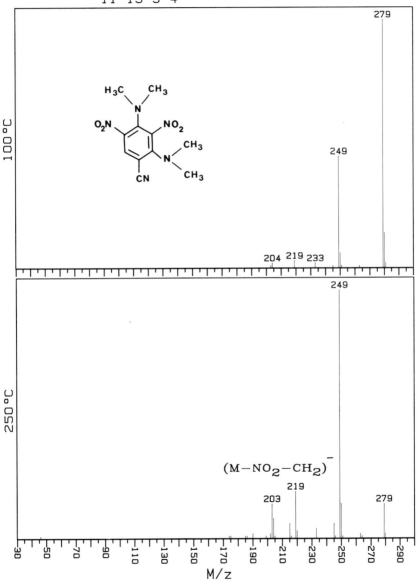

4-Carbamyl-2,6-dinitro-N,N,N,N-tetramethyl-1,3-benzenediamine
CAS No: not avail.
Formula: $C_{11}H_{15}N_5O_5$ MW: 297

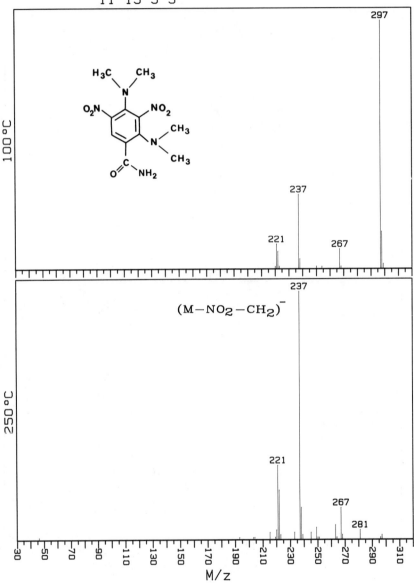

2,6-Dinitro-N,N,N,N-tetraethyl-4-(trifluoromethyl)-1,3-benzenediamine
CAS No: 29263-07-8
Formula: $C_{15}H_{21}F_3N_4O_4$ MW: 378

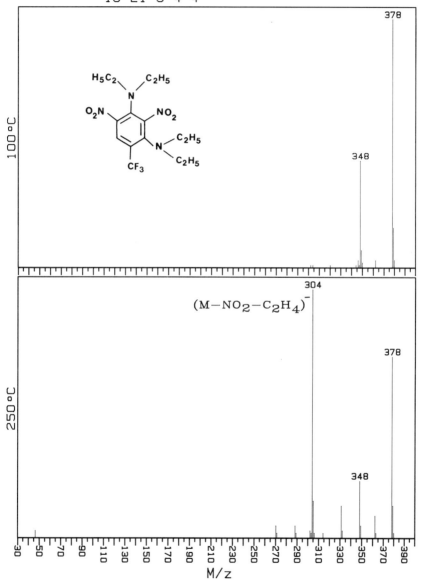

4-Cyano-2,6-dinitro-N,N,N',N'-tetraethyl-1,3-benzenediamine
CAS No: not avail.
Formula: $C_{15}H_{21}N_5O_4$ MW: 335

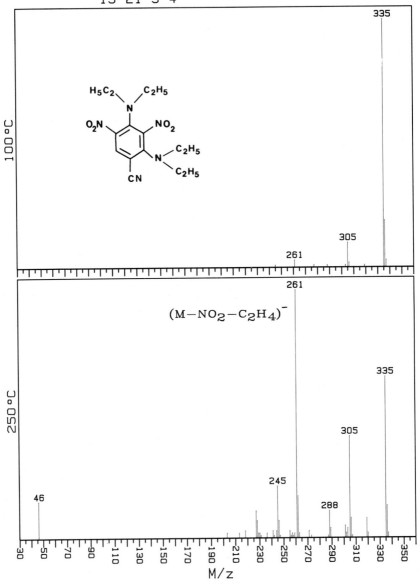

4-Carbamyl-2,6-dinitro-N,N,N,N-tetraethyl-1,3-benzenediamine
CAS No: not avail.
Formula: $C_{15}H_{23}N_5O_5$ MW: 353

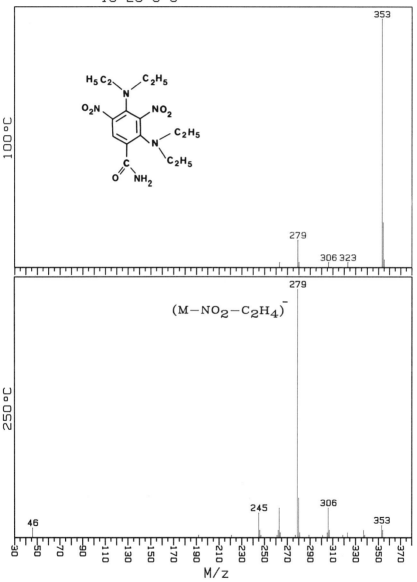

$(M-NO_2-C_2H_4)^-$

4-Cyano-2,6-dinitro-N,N,N,N-tetrapropyl-1,3-benzenediamine
CAS No: not avail.
Formula: $C_{19}H_{29}N_5O_4$ MW: 391

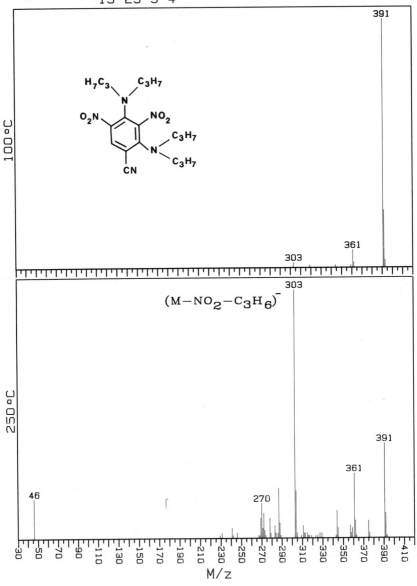

4-Carbamyl-2,6-dinitro-N,N,N,N-tetrapropyl-1,3-benzenediamine
CAS No: not avail.
Formula: $C_{19}H_{31}N_5O_5$ MW: 409

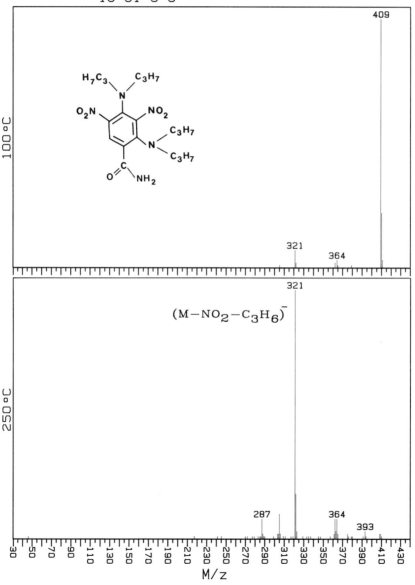

Fluchloralin
CAS No: 33245-39-5
Formula: $C_{12}H_{13}ClF_3N_3O_4$ MW: 355

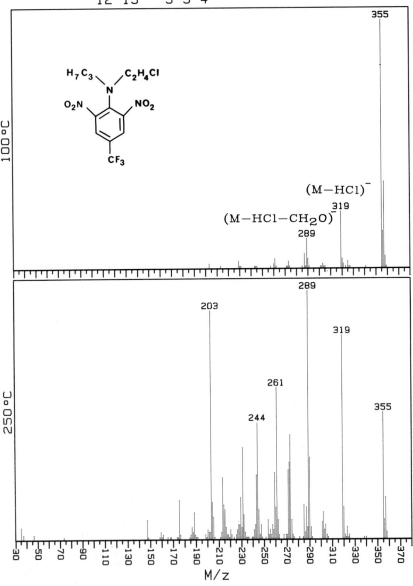

N-(2-Chloroethyl)-2,6-dinitro-4-(trifluoromethyl)aniline
CAS No: 36652-79-6
Formula: $C_9H_7ClF_3N_3O_4$ MW: 313

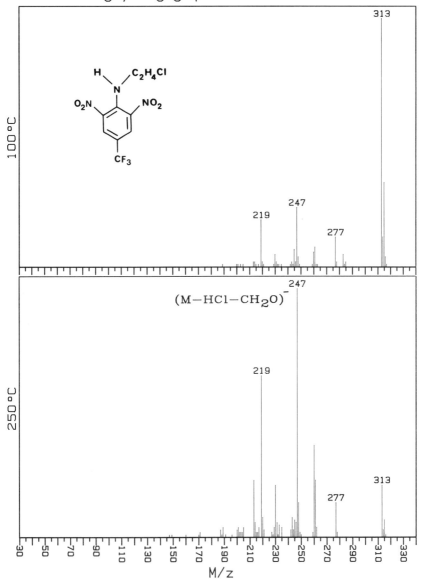

N-(2-Bromoethyl)-2,6-dinitro-4-(trifluoromethyl)aniline
CAS No: not avail.
Formula: $C_9H_7BrF_3N_3O_4$ MW: 357

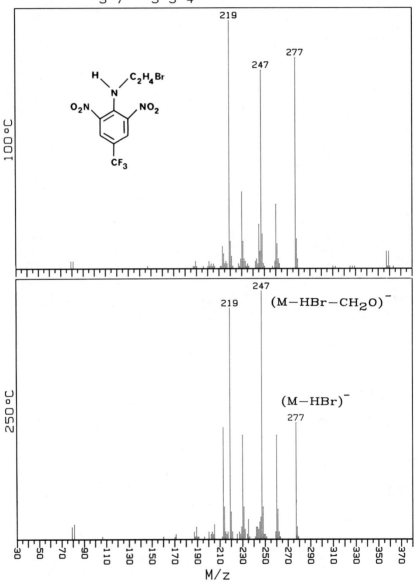

2,6-Dinitro-N-(2-hydroxyethyl)-4-trifluoromethylaniline
CAS No: 17474-02-1
Formula: $C_9H_8F_3N_3O_5$ MW: 295

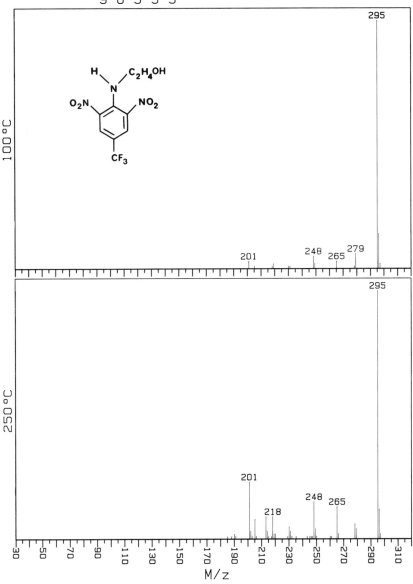

N-(3-Chloropropyl)-2,6-dinitro-4-(trifluoromethyl)aniline
CAS No: 59431-93-5
Formula: $C_{10}H_9N_3O_4ClF_3$ MW: 327

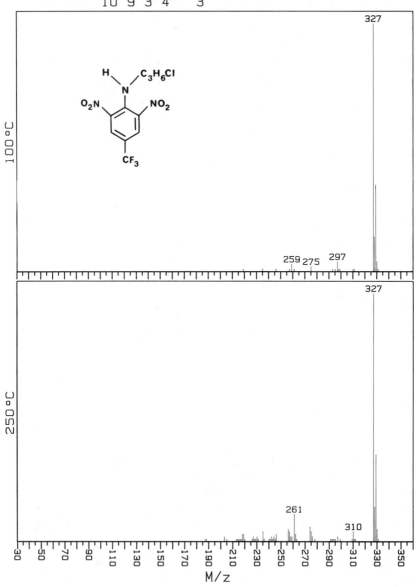

N-(2-Chloroethyl)-2,6-dinitro-4-methoxycarbonaniline
CAS No: not avail.
Formula: $C_{10}H_{10}ClN_3O_6$ MW: 303

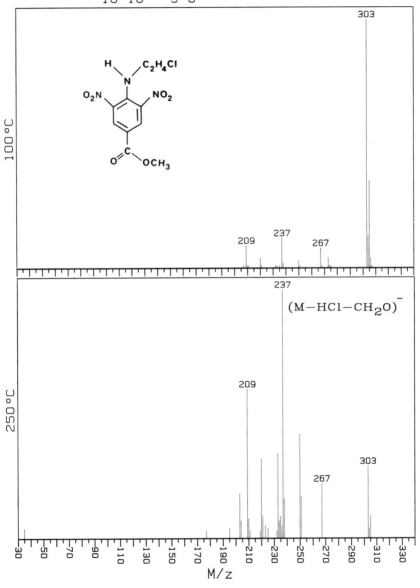

N-(2-Chloroethyl)-4-cyano-2,6-dinitroaniline
CAS No: not avail.
Formula: $C_9H_7ClN_4O_4$ MW: 270

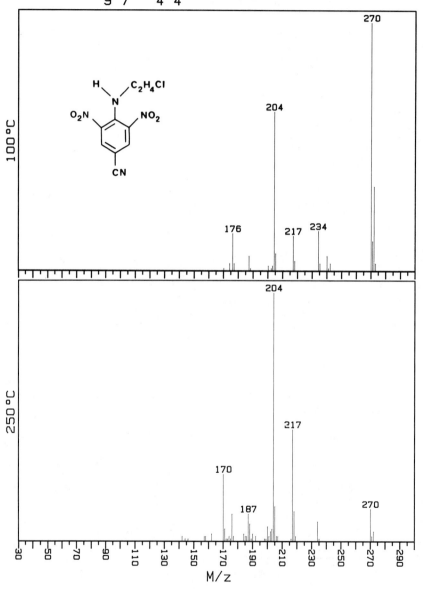

Dimethyl phthalate
CAS No: 131-11-3
Formula: $C_{10}H_{10}O_4$ MW: 194

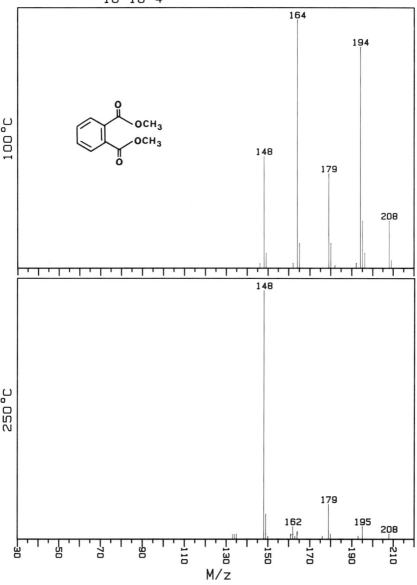

Diethyl phthalate
CAS No: 84-66-2
Formula: $C_{12}H_{14}O_4$ MW: 222

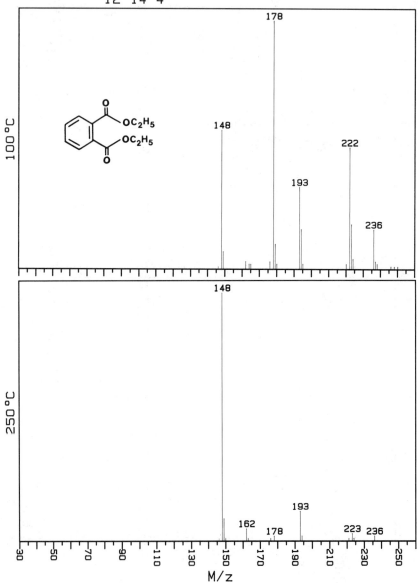

Dibutyl phthalate
CAS No: 84-74-2
Formula: $C_{16}H_{22}O_4$ MW: 278

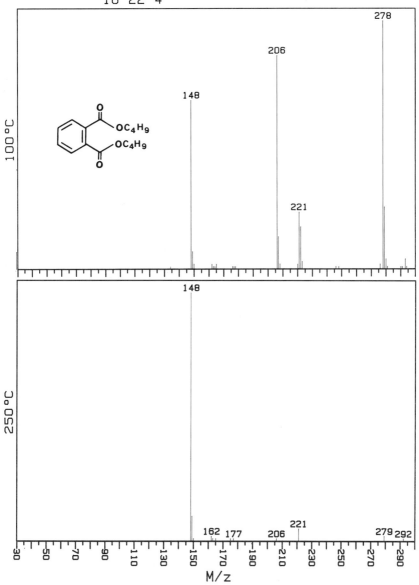

Dioctyl phthalate
CAS No: 117-81-7
Formula: $C_{24}H_{38}O_4$ MW: 390

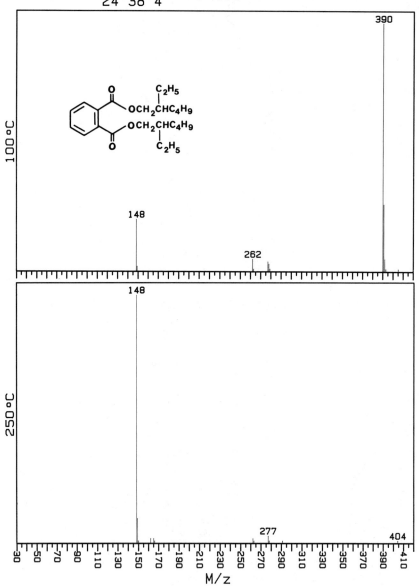

Di-n-octyl phthalate
CAS No: 117-84-0
Formula: $C_{24}H_{38}O_4$ MW: 390

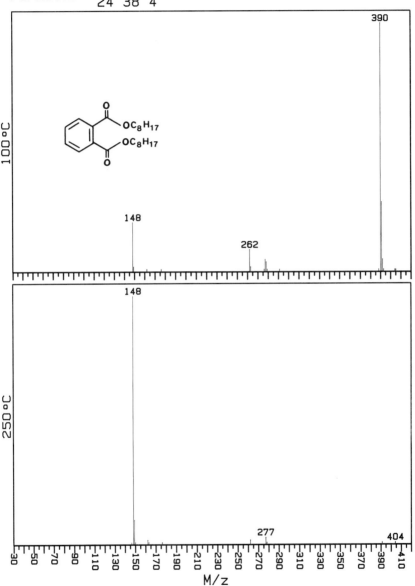

Dinonyl phthalate
CAS No: 84-76-4
Formula: $C_{26}H_{42}O_4$ MW: 418

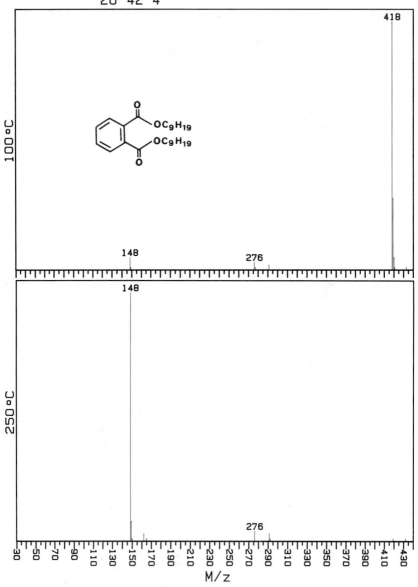

Diphenyl phthalate
CAS No: 84-62-8
Formula: $C_{20}H_{14}O_4$ MW: 318

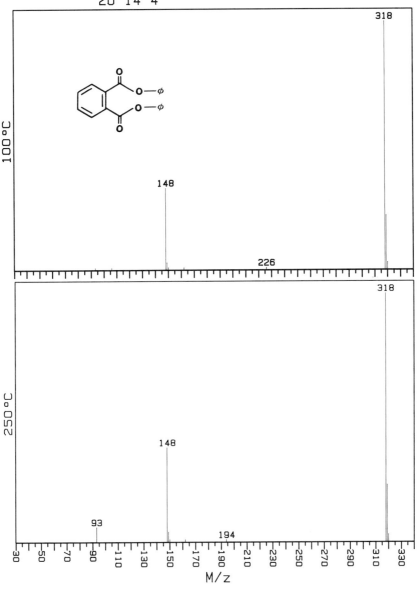

Butyl benzyl phthalate
CAS No: 85-68-7
Formula: $C_{19}H_{20}O_4$ MW: 312

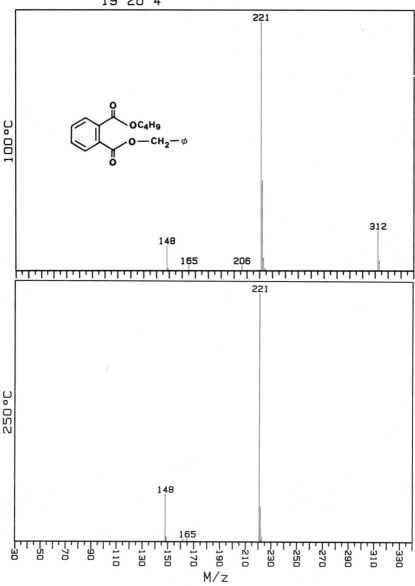

Dimethyl terephthalate
CAS No: not avail.
Formula: $C_{10}H_{10}O_4$ MW: 194

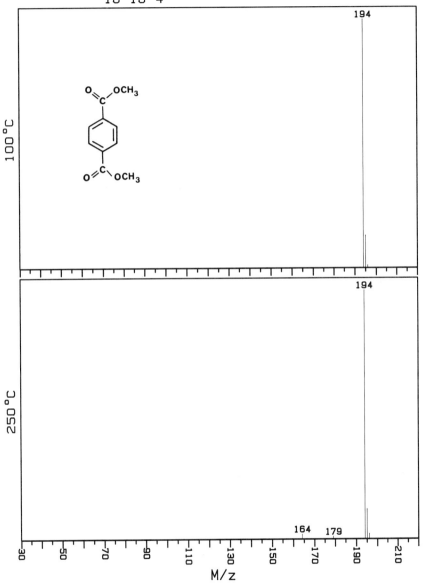

Dacthal
CAS No: 1861-32-1
Formula: $C_{10}H_6Cl_4O_4$ MW: 330

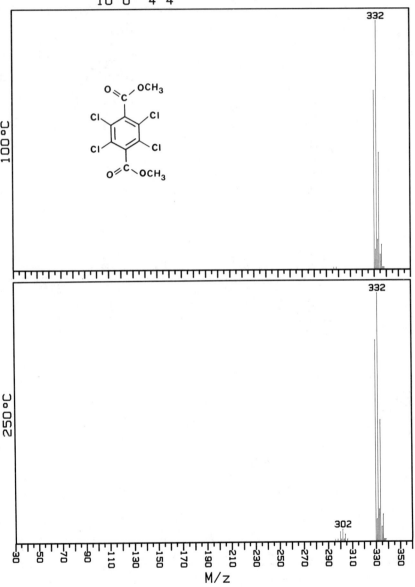

Diphenylfulvene
CAS No: 2175-90-8
Formula: $C_{18}H_{14}$ MW: 230

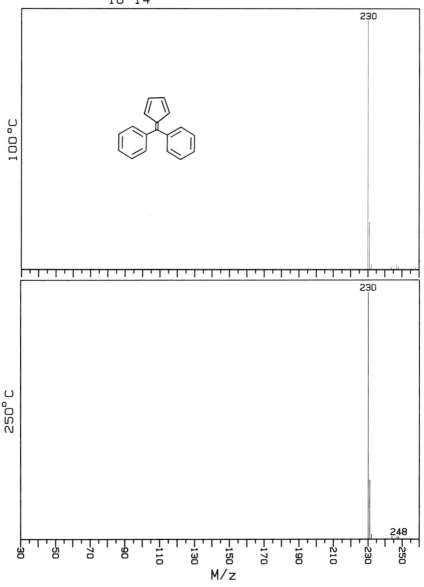

Fluoranthene
CAS No: 206-44-0
Formula: $C_{16}H_{10}$ MW: 202

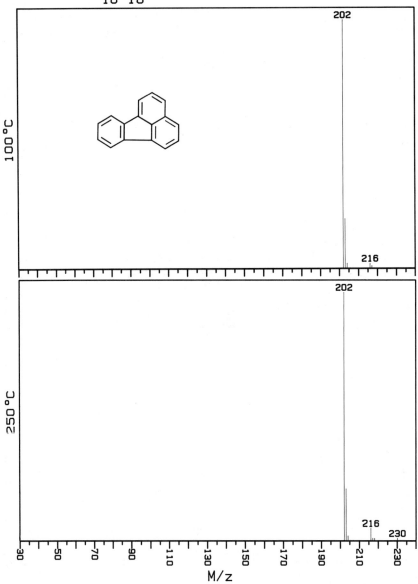

Benzo[a]pyrene
CAS No: 50-32-8
Formula: $C_{20}H_{12}$ MW: 252

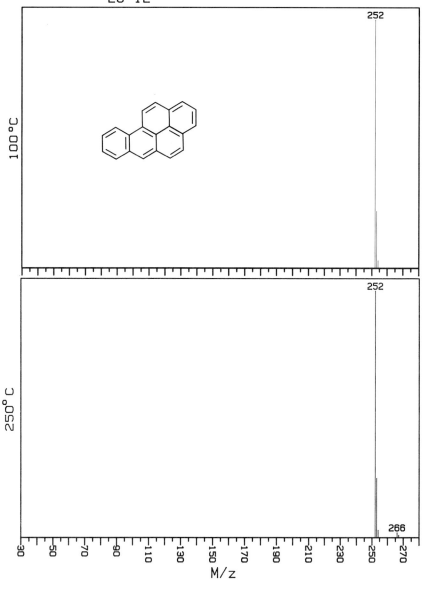

Benzo[ghi]perylene
CAS No: 191-24-2
Formula: $C_{22}H_{12}$ MW: 276

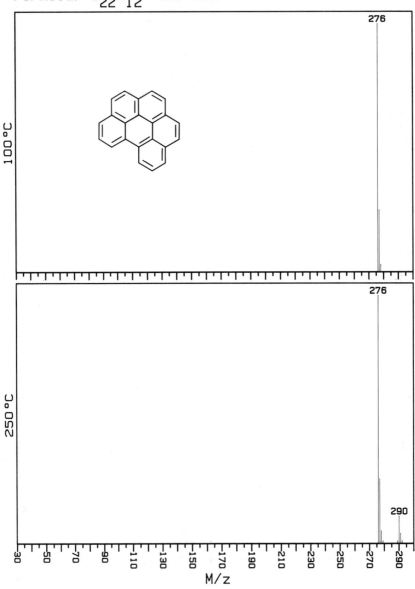

100 °C

276

250 °C

276

290

M/z

1-Nitronaphthalene
CAS No: 86-57-7
Formula: $C_{10}H_7NO_2$ MW: 173

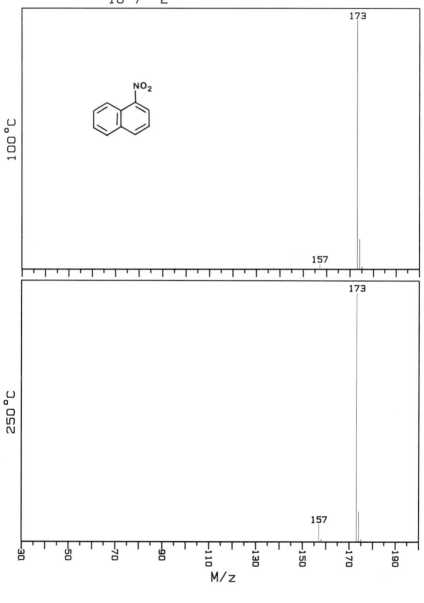

1-Cyanonaphthalene
CAS No: 86-53-3
Formula: $C_{11}H_7N$ MW: 153

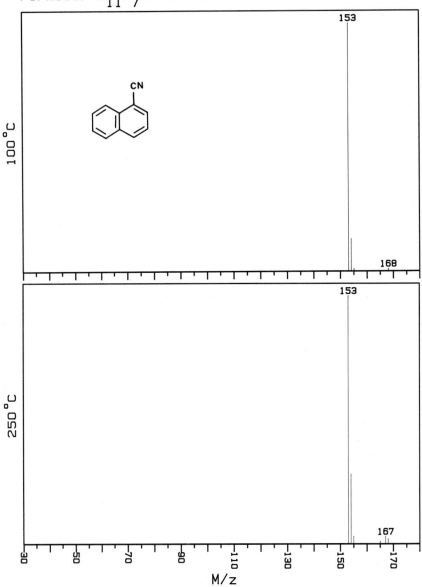

1-Naphthaldehyde
CAS No: 66-77-3
Formula: $C_{11}H_8O$ MW: 156

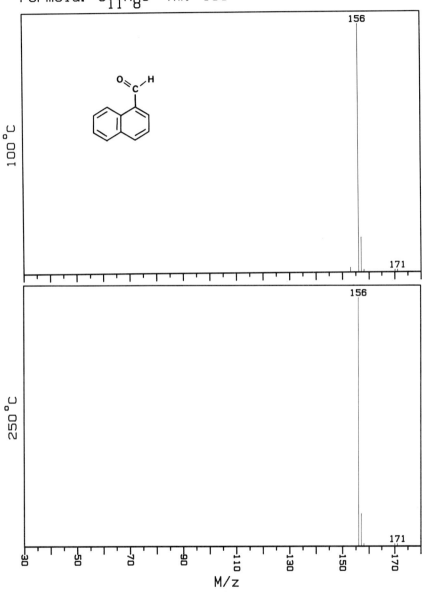

1-Bromonaphthalene
CAS No: 90-11-9
Formula: C$_{10}$H$_7$Br   MW: 206

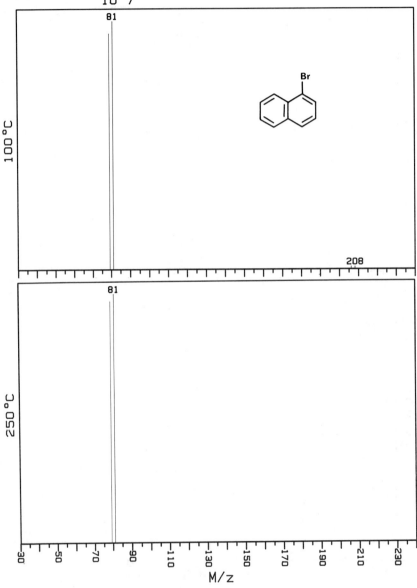

1-Chloronaphthalene
CAS No: 90-13-1
Formula: $C_{10}H_7Cl$ MW: 162

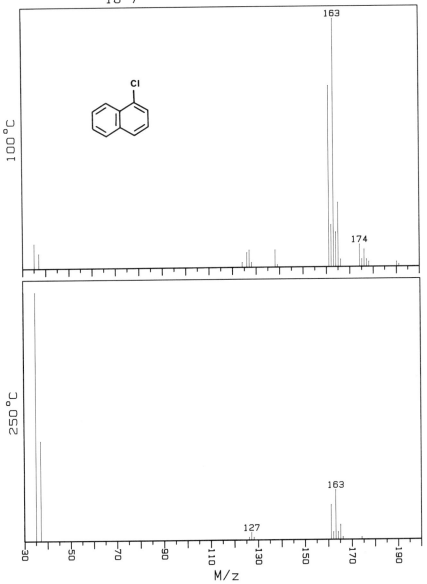

1,8-Dichloronaphthalene
CAS No: 2050-74-0
Formula: $C_{10}H_6Cl_2$ MW: 196

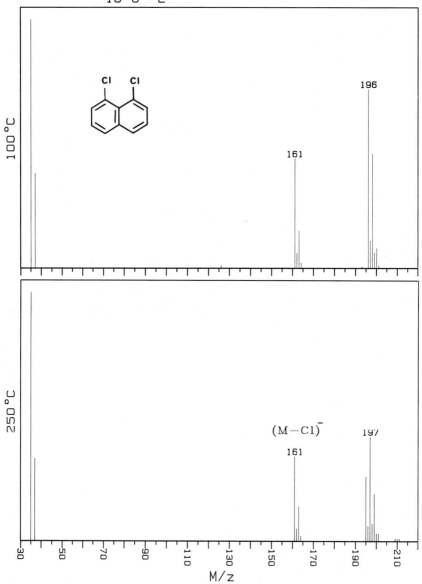

Trichloronaphthalene (Halowax 1014)
CAS No: 12616-36-3
Formula: C$_{10}$H$_7$Cl$_3$ MW: 230

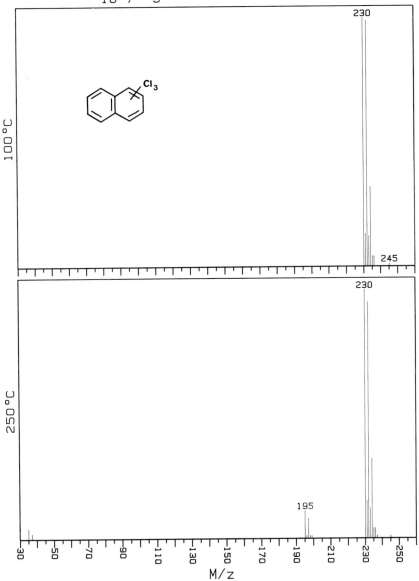

Tetrachloronaphthalene (Halowax 1014)
CAS No: 12616-36-3
Formula: $C_{10}H_4Cl_4$ MW: 264

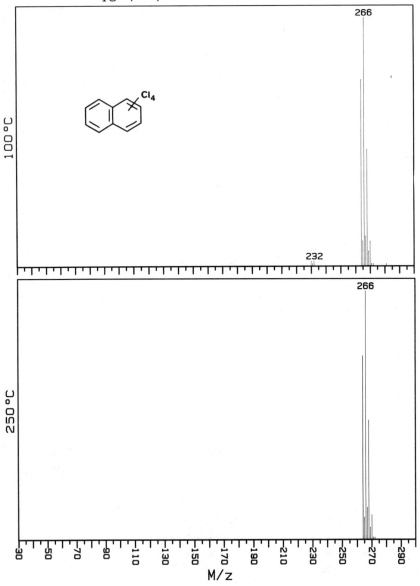

Pentachloronaphthalene (Halowax 1014)
CAS No: 12616-36-3
Formula: $C_{10}H_5Cl_3$ MW: 298

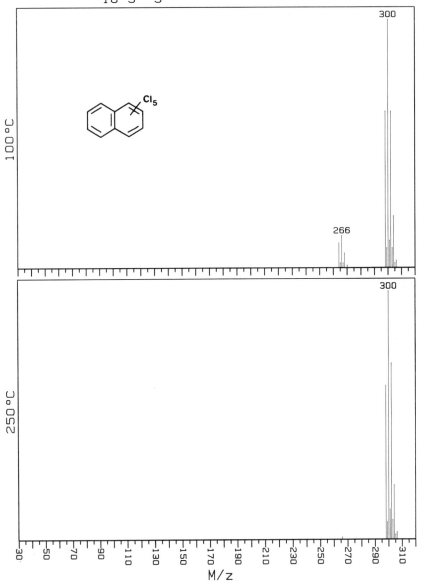

Hexachloronaphthalene (Halowax 1014)
CAS No: 12616-36-3
Formula: $C_{10}H_2Cl_6$ MW: 332

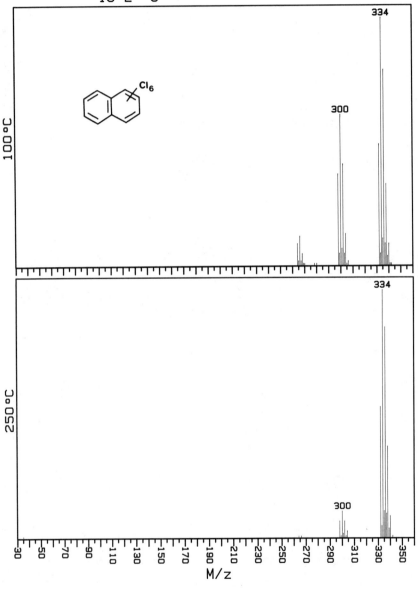

Heptachloronaphthalene (Halowax 1014)
CAS No: 12616-36-3
Formula: $C_{10}HCl_7$ MW: 366

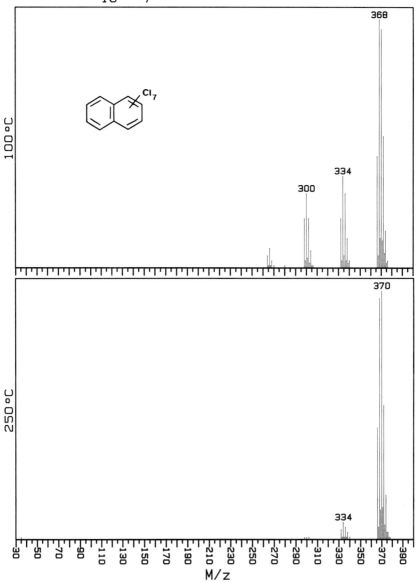

Octachloronaphthalene
CAS No: 2234-13-1
Formula: $C_{10}Cl_8$ MW: 400

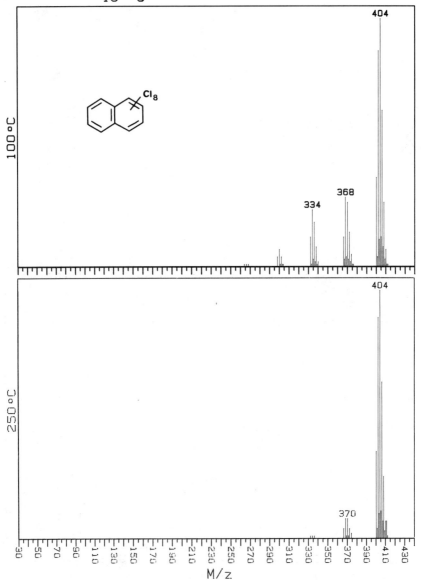

Octafluoronaphthalene
CAS No: 313-72-4
Formula: $C_{10}F_8$ MW: 272

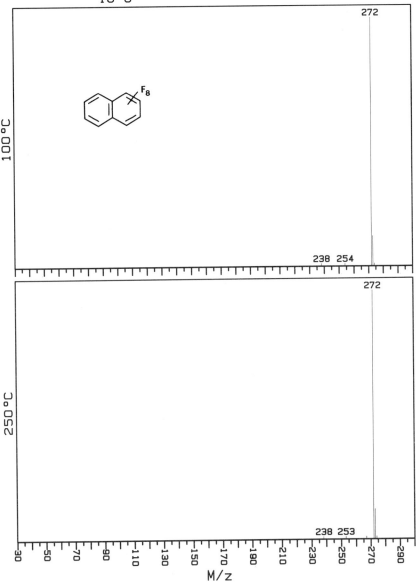

Dichlone
CAS No: 117-80-6
Formula: $C_{10}H_4Cl_2O_2$ MW: 226

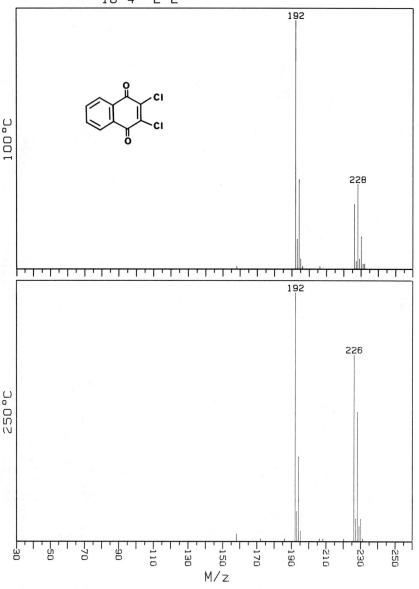

2,4-Dichloronaphthol
CAS No: 2050-76-2
Formula: $C_{10}H_6Cl_2O$ MW: 212

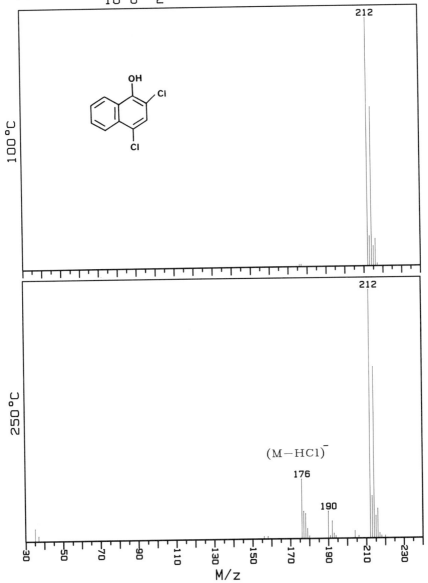

4,4′-Dichlorobiphenyl
CAS No: 2050-68-6
Formula: $C_{12}H_8Cl_2$ MW: 222

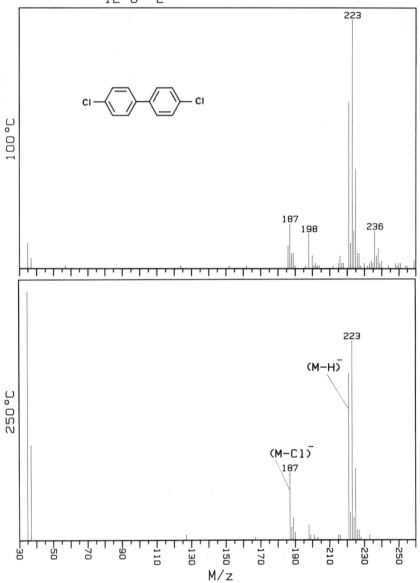

2,4'-Dichlorobiphenyl
CAS No: 34883-43-7
Formula: $C_{12}H_8Cl_2$ MW: 222

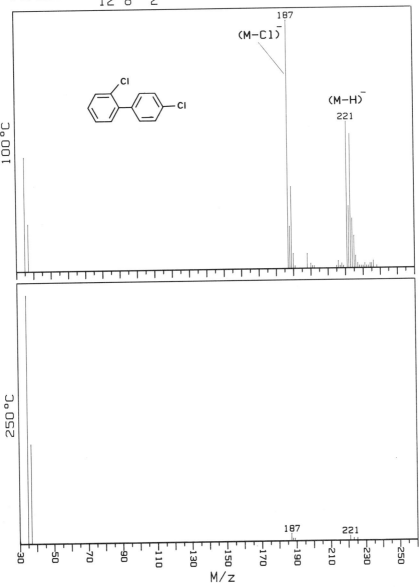

2,2'-Dichlorobiphenyl
CAS No: 13029-08-8
Formula: $C_{12}H_8Cl_2$ MW: 222

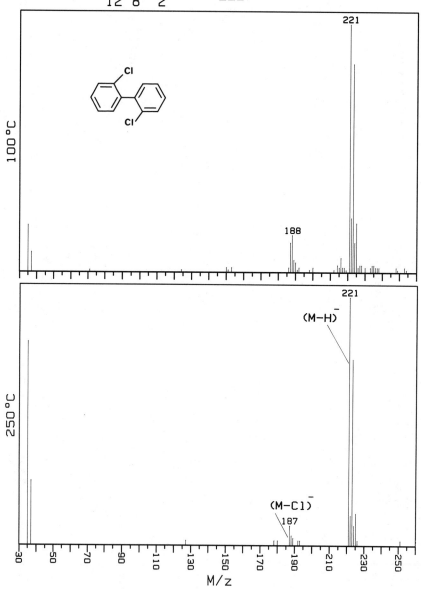

2,2',5-Trichlorobiphenyl
CAS No: 37680-65-2
Formula: $C_{12}H_7Cl_3$ MW: 256

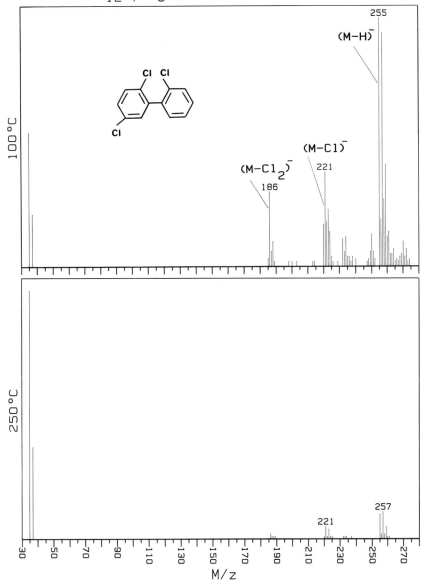

2,2',5,5'-Tetrachlorobiphenyl
CAS No: 35693-99-3
Formula: $C_{12}H_6Cl_4$ MW: 290

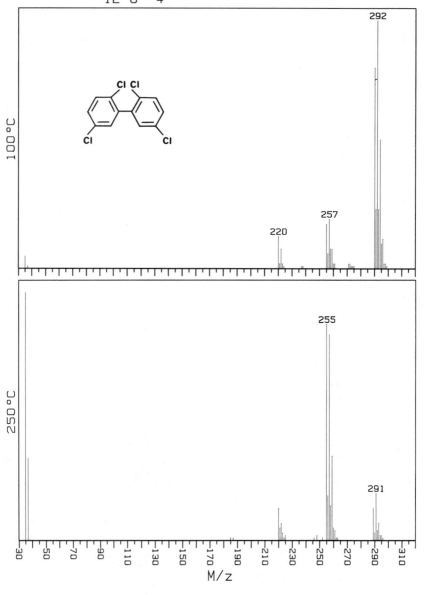

2,2',4,5,5'-Pentachlorobiphenyl
CAS No: 37680-73-2
Formula: $C_{12}H_5Cl_5$ MW: 324

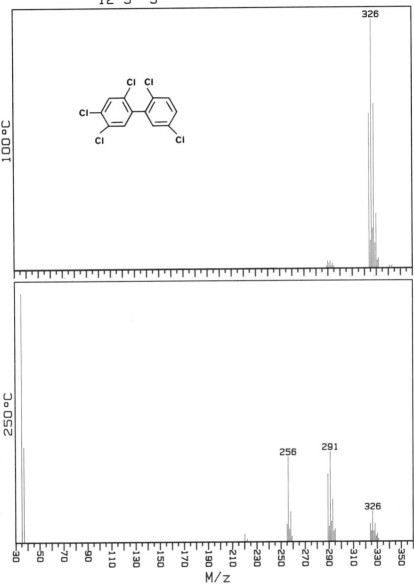

2, 3, 4, 5, 6-Pentachlorobiphenyl
CAS No: 18259-05-7
Formula: $C_{12}H_5Cl_5$ MW: 324

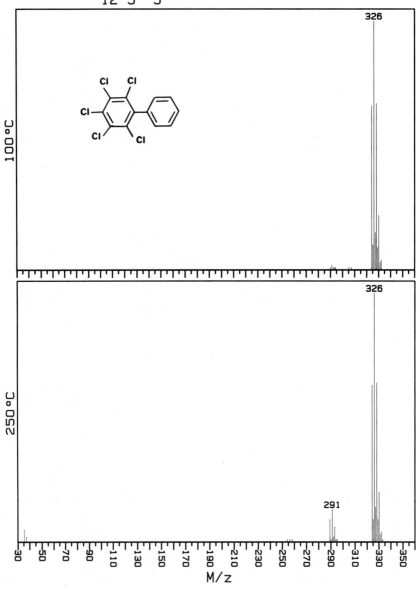

2,2',4,4',5,5'-Hexachlorobiphenyl
CAS No: 52663-72-6
Formula: $C_{12}H_4Cl_6$ MW: 358

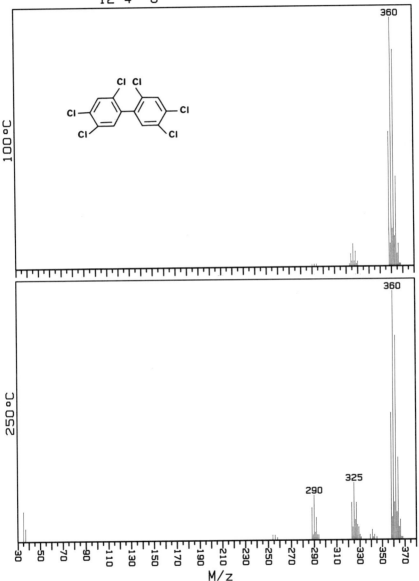

2, 2', 3, 4, 4', 5, 5'-Heptachlorobiphenyl
CAS No: 35065-29-3
Formula: $C_{12}H_3Cl_7$ MW: 392

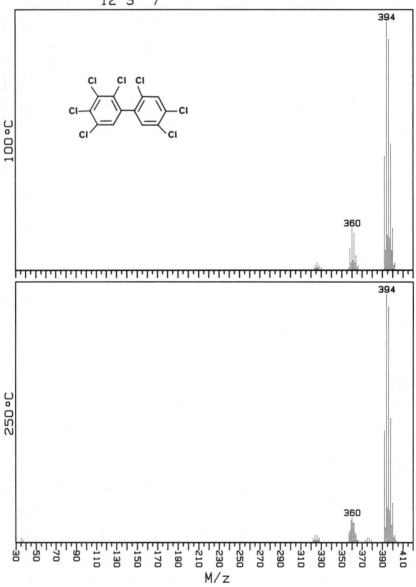

2, 2', 3, 3', 4, 4', 5, 5'-Octachlorobiphenyl
CAS No: 35694-08-7
Formula: $C_{12}H_2Cl_8$ MW: 426

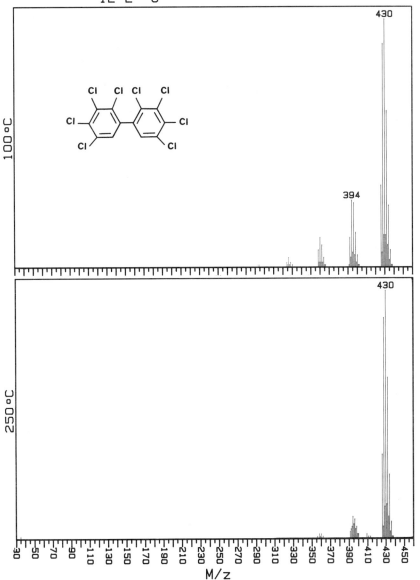

2, 2′, 3, 3′, 4, 4′, 5, 5′, 6-Nonachlorobiphenyl
CAS No: 40186-72-9
Formula: $C_{12}HCl_9$ MW: 460

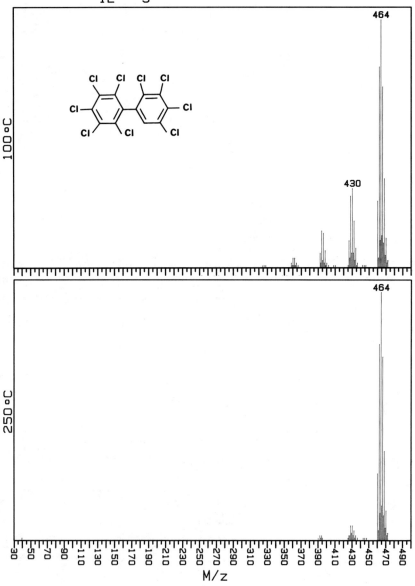

Decachlorobiphenyl
CAS No: 2051-24-3
Formula: $C_{12}Cl_{10}$ MW: 494

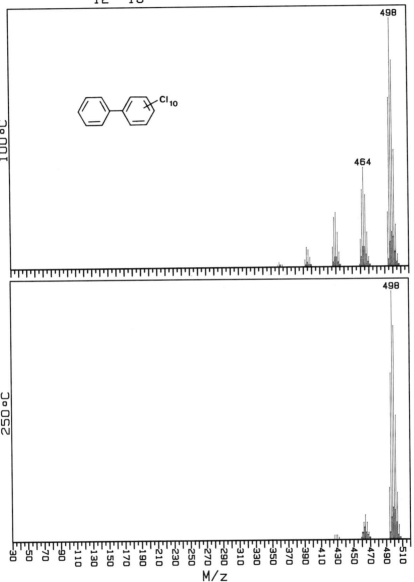

Decafluorobiphenyl
CAS No: 434-90-2
Formula: $C_{12}F_{10}$ MW: 334

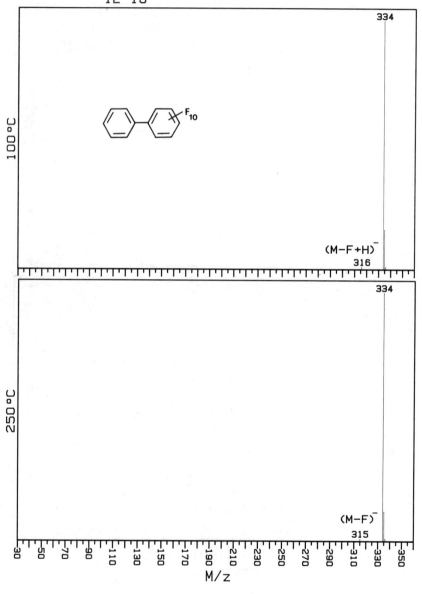

4,4'-Dibromobiphenyl
CAS No: 92-86-4
Formula: $C_{12}H_8Br_2$ MW: 310

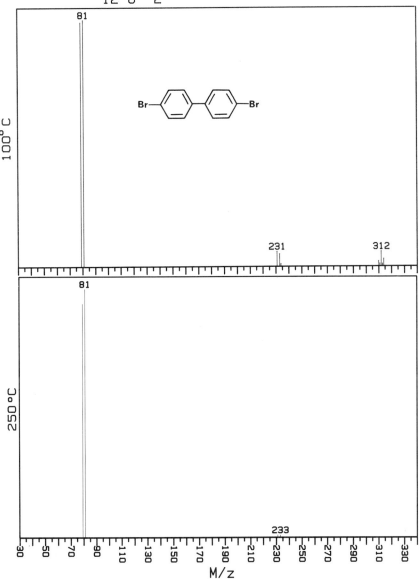

2, 4, 5-Tribromobiphenyl
CAS No: 51202-79-0
Formula: $C_{12}H_7Br_3$ MW: 388

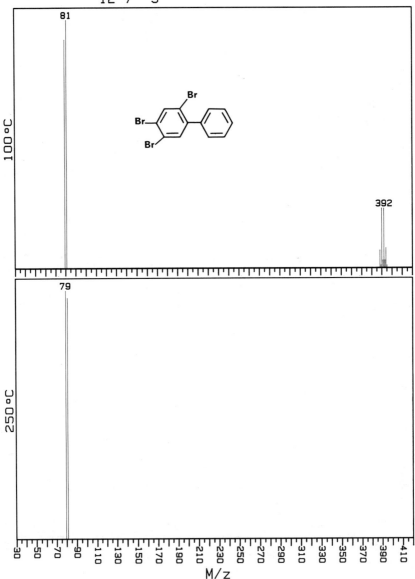

2,2',5,5'-Tetrabromobiphenyl
CAS No: 59080-37-4
Formula: $C_{12}H_6Br_4$ MW: 466

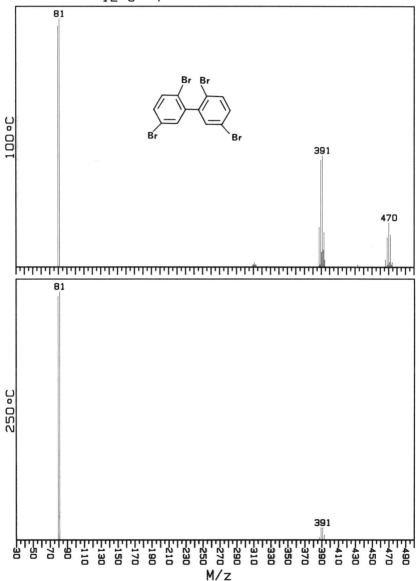

2, 2', 4, 5, 5'-Pentabromobiphenyl
CAS No: 67888-96-4
Formula: $C_{12}H_5Br_5$ MW: 544

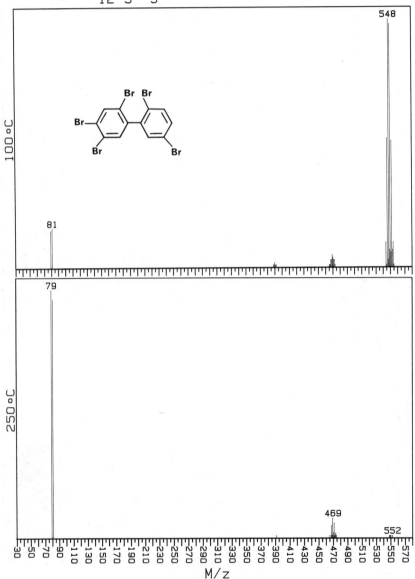

2,2',4,4',5,5'-Hexabromobiphenyl
CAS No: 59080-40-9
Formula: $C_{12}H_4Br_6$ MW: 622

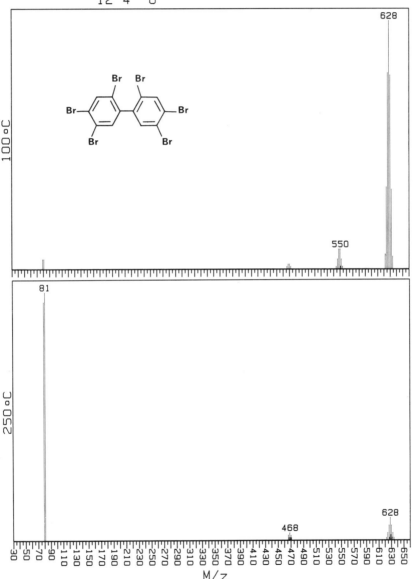

2,2'-Dinitrobiphenyl
CAS No: 2436-96-6
Formula: $C_{12}H_8N_2O_4$ MW: 244

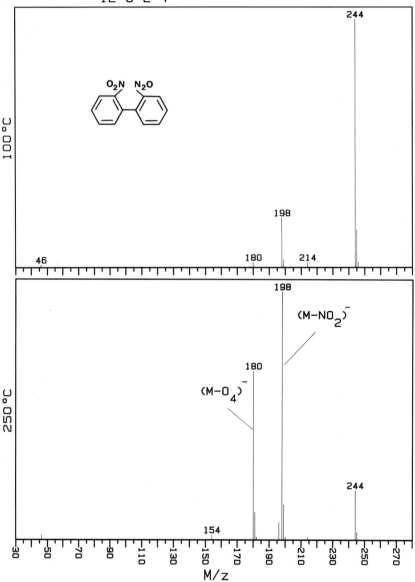

2,2'-Dinitrobibenzyl
CAS No: 16968-19-7
Formula: $C_{14}H_{12}N_2O_4$ MW: 272

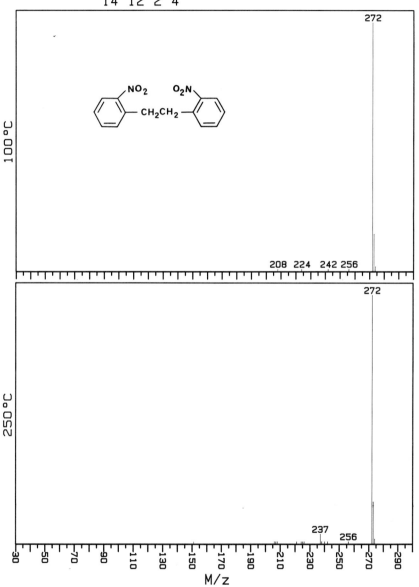

4,4'-Dinitrobibenzyl
CAS No: 736-30-1
Formula: $C_{14}H_{12}N_2O_4$ MW: 272

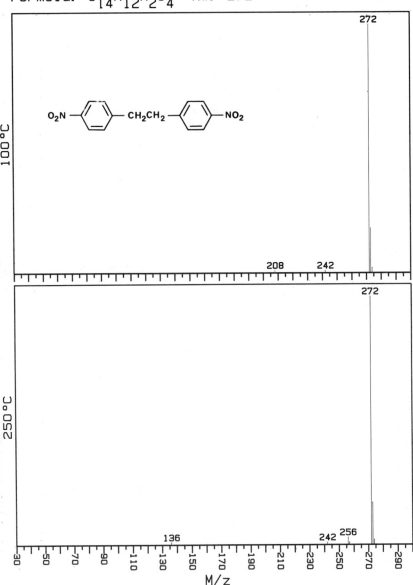

2′,3′,4′,5′-Tetrachloro-4-biphenyol
CAS No: 67651-34-7
Formula: C$_{12}$H$_6$Cl$_4$O   MW: 306

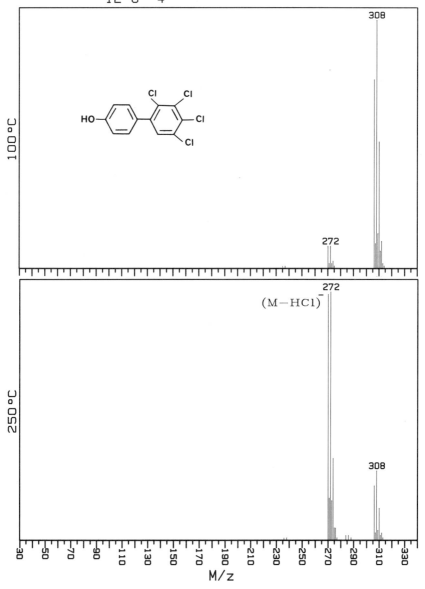

2',3',4',5'-Tetrachloro-3-biphenyol
CAS No: 67651-37-0
Formula: $C_{12}H_6Cl_4O$ MW: 306

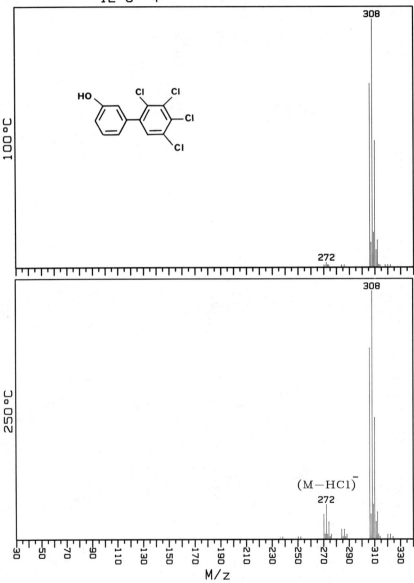

3,3′,5,5′-Tetrachloro-4,4′-biphenyldiol
CAS No: 13049-13-3
Formula: $C_{12}H_6Cl_4O_2$ MW: 322

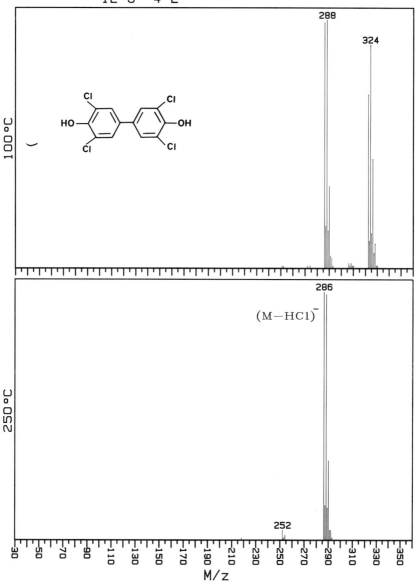

2′,3,3′,4′,5′-Pentachloro-2-biphenyol
CAS No: 67651-35-8
Formula: $C_{12}H_5Cl_5O$ MW: 340

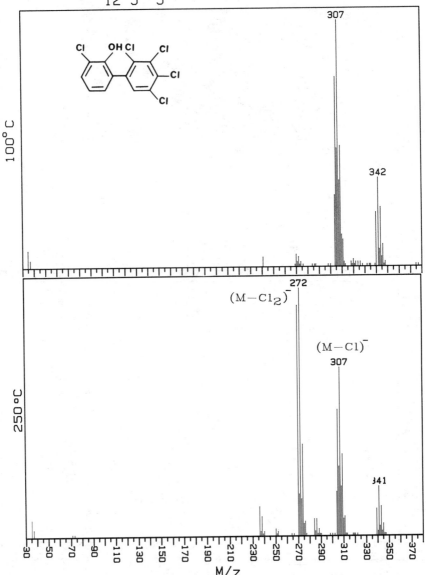

2′,3′,4′,5,5′-Pentachloro-2-biphenylol
CAS No: 67651-36-9
Formula: $C_{12}H_5Cl_5O$ MW: 340

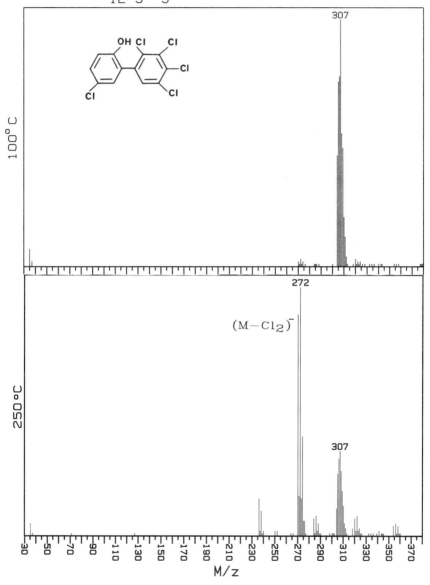

2,4'',5-Trichloro-p-terphenyl
CAS No: 61576-93-0
Formula: $C_{18}H_{11}Cl_3$ MW: 332

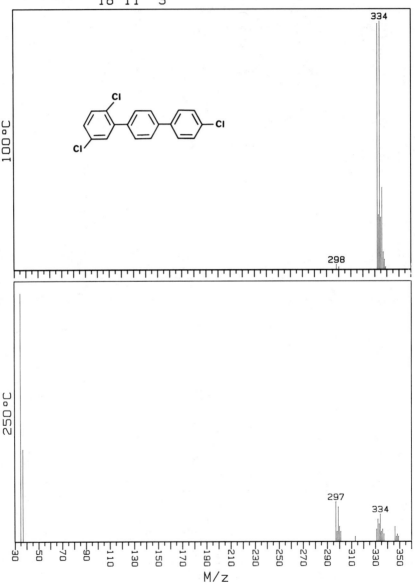

2,4,4'',6-Tetrachloro-p-terphenyl
CAS No: 61576-97-4
Formula: $C_{18}H_{10}Cl_4$ MW: 366

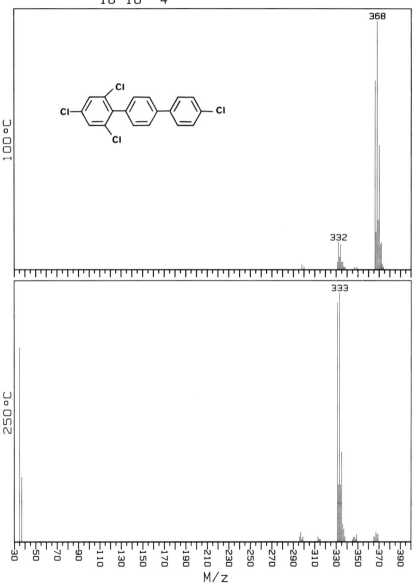

2,2',4-Trichlorodiphenyl ether
CAS No: 68914-97-6
Formula: $C_{12}H_7Cl_3O$ MW: 272

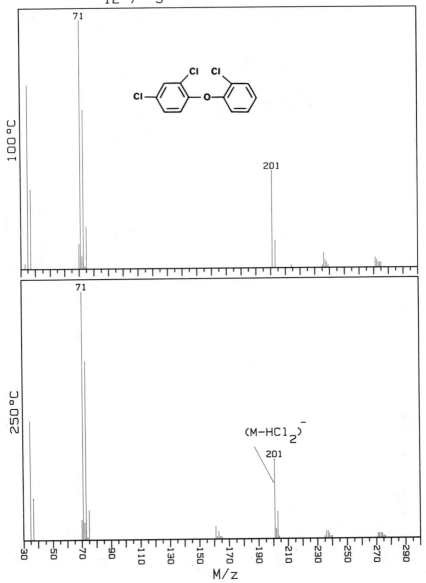

2, 4, 4', 5-Tetrachlorodiphenyl ether
CAS No: 61328-45-8
Formula: $C_{12}H_6Cl_4O$ MW: 306

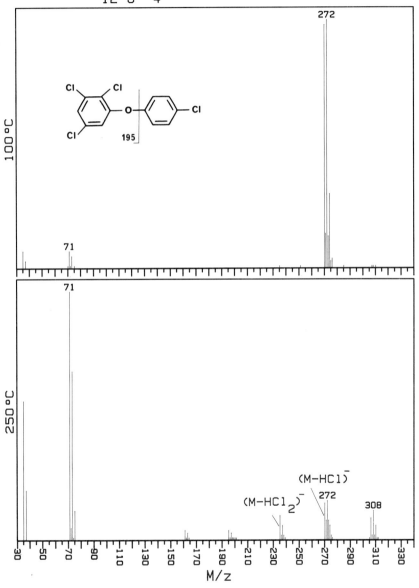

2',3,4,4'-Tetrachlorodiphenyl ether
CAS No: 61328-46-9
Formula: $C_{12}H_6Cl_4O$ MW: 306

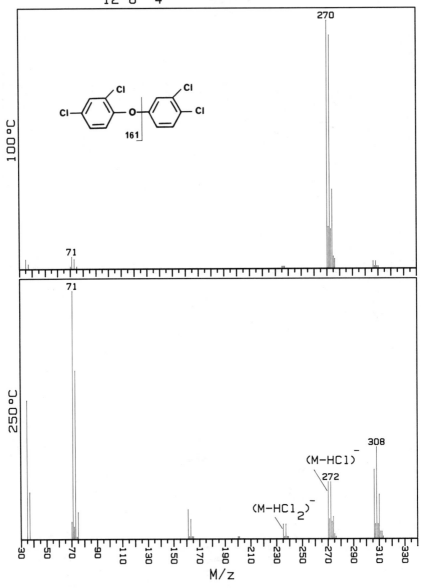

3,3',4,4'-Tetrachlorodiphenyl ether
CAS No: 56348-72-2
Formula: $C_{12}H_6Cl_4O$ MW: 306

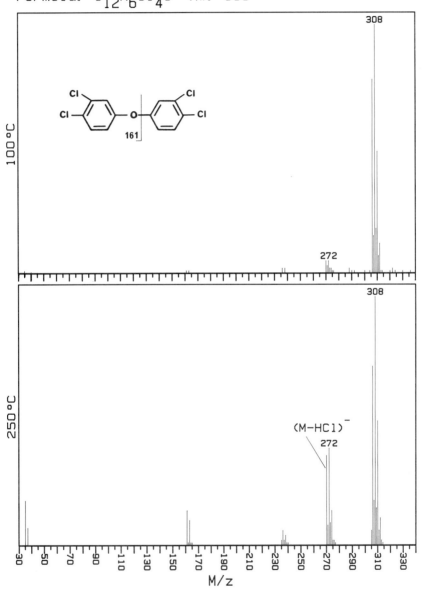

2,2',4,4',5-Pentachlorodiphenyl ether
CAS No: 60123-64-0
Formula: $C_{12}H_5Cl_5O$ MW: 340

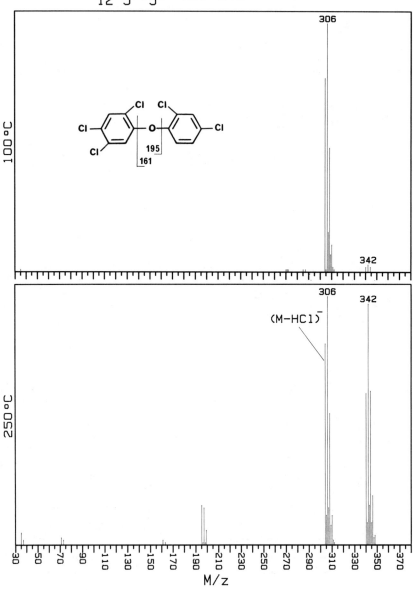

3, 3', 4, 4', 5-Pentachlorodiphenyl ether
CAS No: 94339-59-0
Formula: $C_{12}H_5Cl_5O$ MW: 340

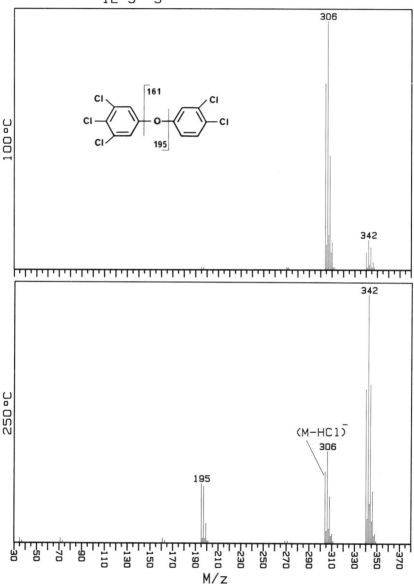

Decachlorodiphenyl ether
CAS No: 31710-30-2
Formula: $C_{12}Cl_{10}O$ MW: 510

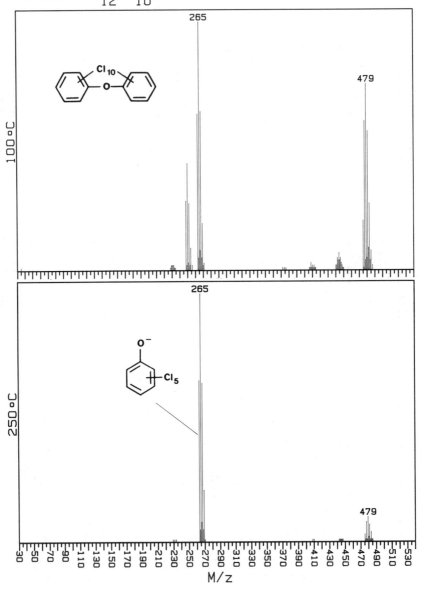

Tetrasul
CAS No: 2227-13-6
Formula: $C_{12}H_6Cl_4S$ MW: 322

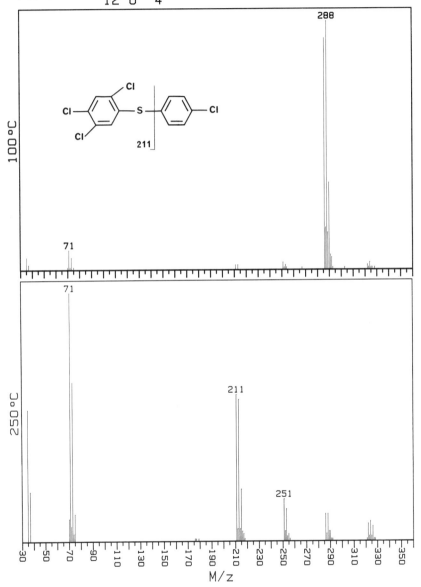

Tetrabromodiphenyl ether (Bromkal 70-5 DE)
CAS No: 40088-47-9
Formula: $C_{12}H_6Br_4O$ MW: 482

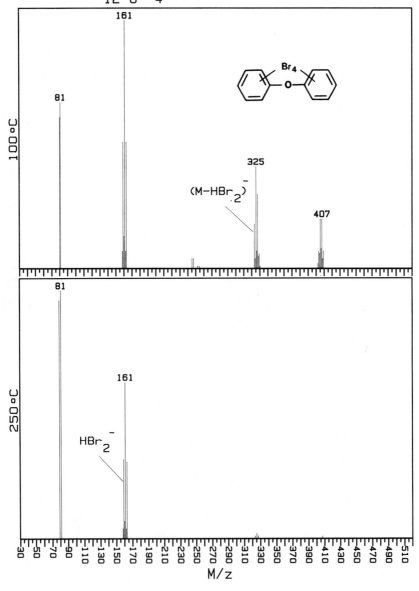

Pentabromodiphenyl ether (Bromkal 70-5 DE)
CAS No: 32534-81-9
Formula: $C_{12}H_5Br_5O$ MW: 560

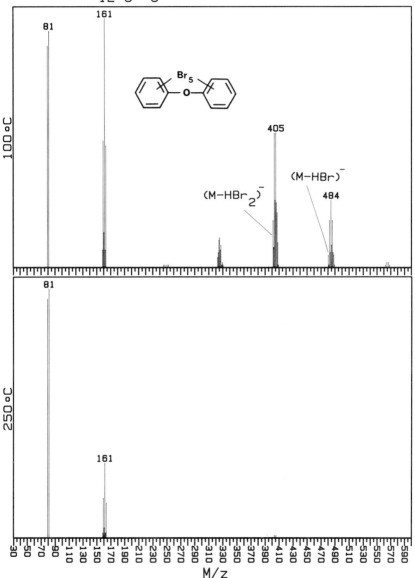

Hexabromodiphenyl ether (Bromkal 70-5 DE)
CAS No: 36483-60-0
Formula: $C_{12}H_4Br_6O$ MW: 638

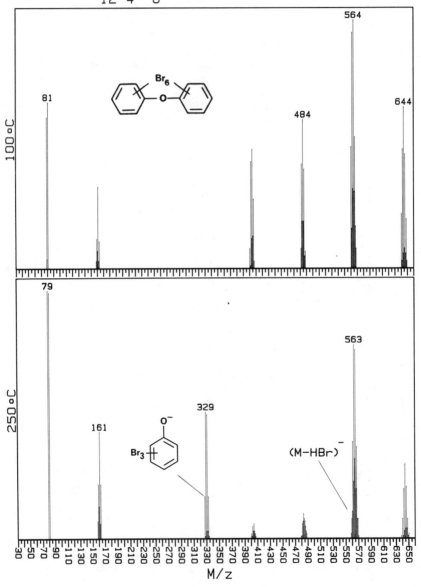

2,4,4′-Trichloro-2′-hydroxydiphenyl ether
CAS No: 3380-34-5
Formula: $C_{12}H_7Cl_3O_2$ MW: 288

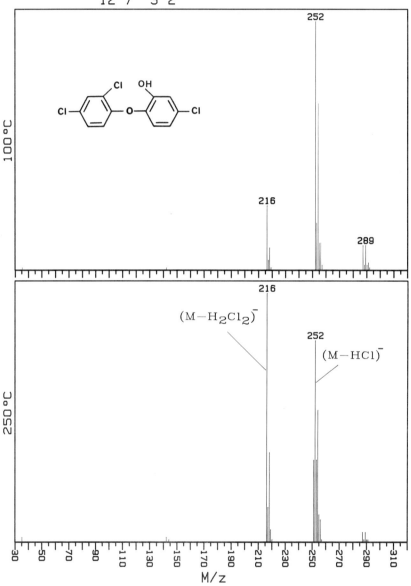

Nitrofen
CAS No: 1836-75-5
Formula: $C_{12}H_7Cl_2NO_3$ MW: 283

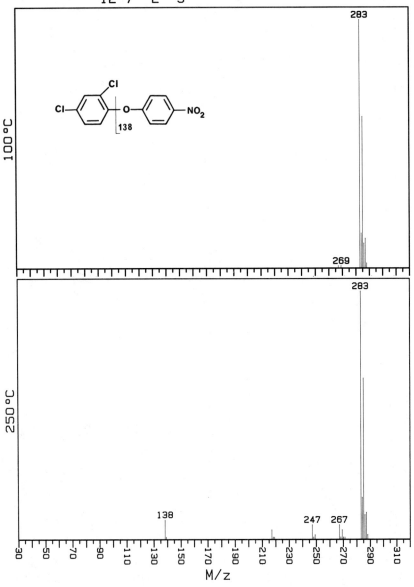

Oxyfluorfen
CAS No: 42874-03-3
Formula: $C_{15}H_{11}ClF_3NO_4$ MW: 361

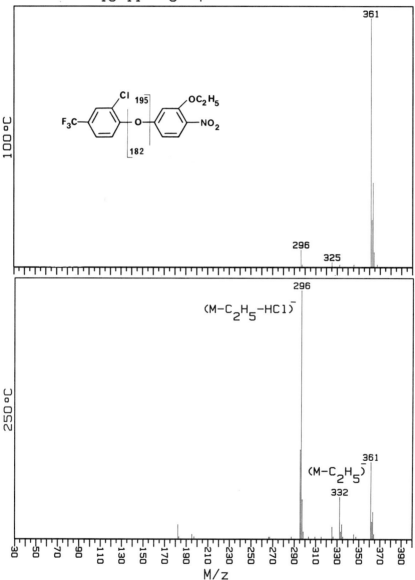

$(M-C_2H_5-HCl)^-$

$(M-C_2H_5)^-$

M/z

Bifenox
CAS No: 42576-02-3
Formula: $C_{14}H_9Cl_2NO_5$ MW: 341

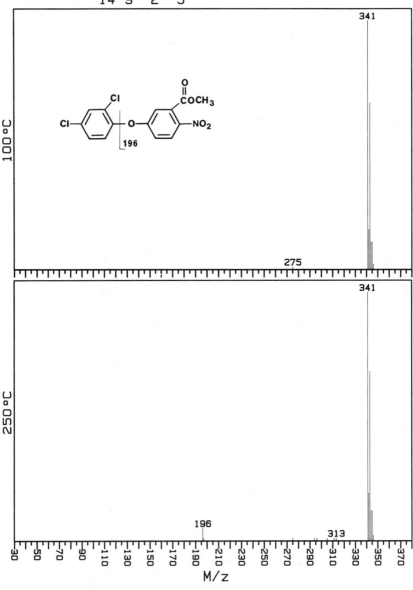

Diclofop-methyl
CAS No: 51338-27-3
Formula: $C_{16}H_{14}Cl_2O_4$ MW: 340

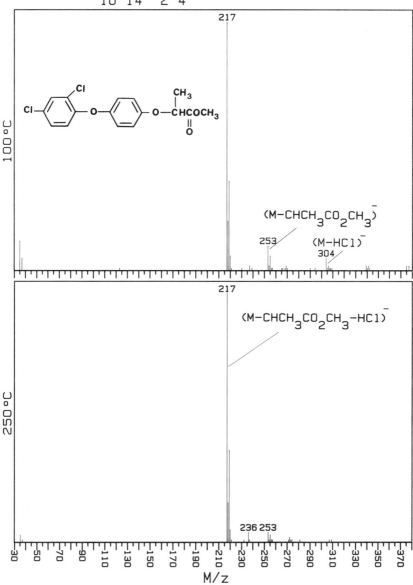

1,2-Bis(4-nitrophenoxy)ethane
CAS No: 14467-69-7
Formula: $C_{14}H_{12}N_2O_6$ MW: 304

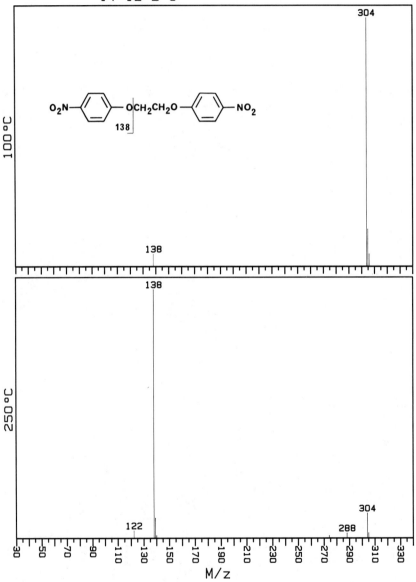

1,2-Bis(2-nitrophenoxy)ethane
CAS No: 51661-19-9
Formula: $C_{14}H_{12}N_2O_6$ MW: 304

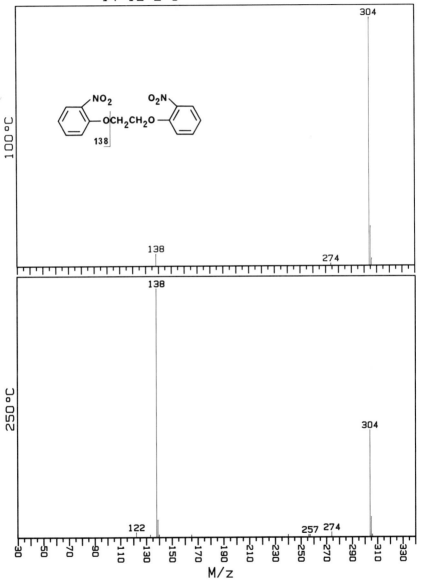

1,2,3,4-Tetrachlorodibenzo-p-dioxin
CAS No: 30746-58-8
Formula: $C_{12}H_4Cl_4O_2$ MW: 320

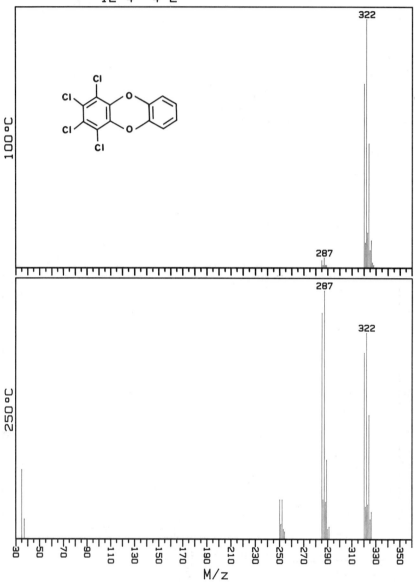

1, 2, 3, 4, 7-Pentachlorodibenzo-p-dioxin
CAS No: 39227-61-7
Formula: $C_{12}H_3Cl_5O_2$ MW: 354

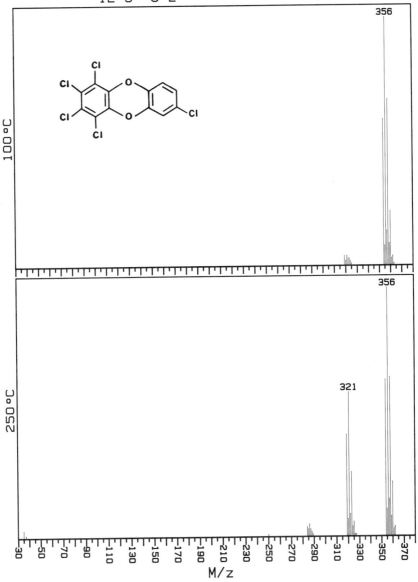

1, 2, 3, 4, 7, 8-Hexachlorodibenzo-p-dioxin
CAS No: 39227-28-6
Formula: $C_{12}H_2Cl_6O_2$ MW: 388

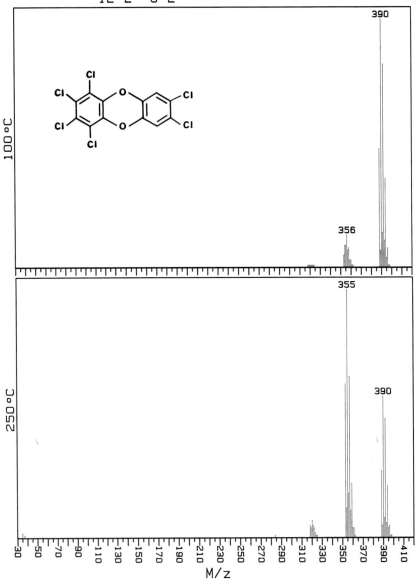

1, 2, 3, 4, 6, 7, 8-Heptachlorodibenzo-p-dioxin
CAS No: 35822-46-9
Formula: $C_{12}HCl_7O_2$ MW: 422

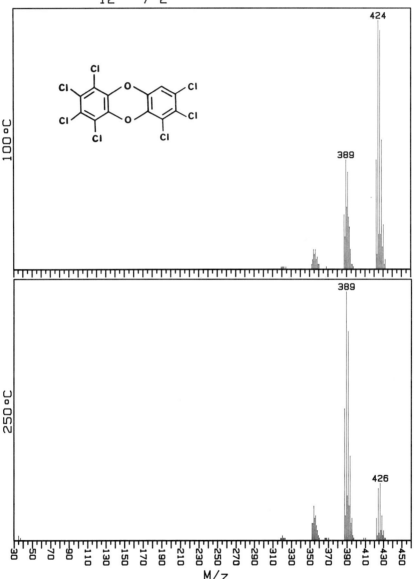

Octachlorodibenzo-p-dioxin
CAS No: 3268-87-9
Formula: $C_{12}Cl_8O_2$ MW: 456

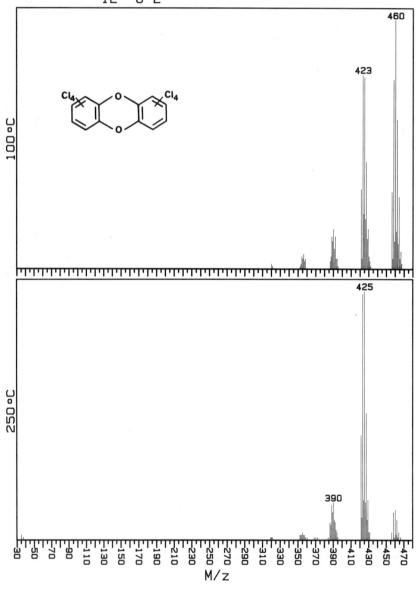

1, 2, 7, 8-Tetrachlorodibenzofuran
CAS No: 58802-20-3
Formula: $C_{12}H_4Cl_4O$ MW: 304

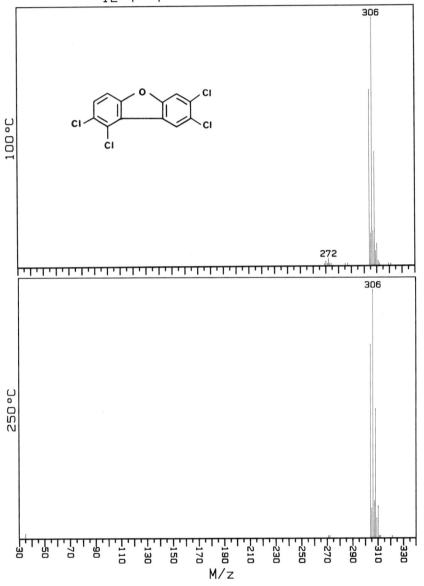

1, 2, 3, 8, 9-Pentachlorodibenzofuran
CAS No: 83704-54-5
Formula: $C_{12}H_3Cl_5O$ MW: 338

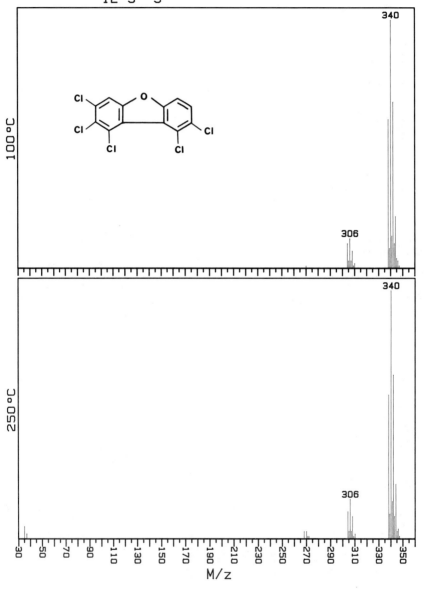

1, 2, 3, 4, 8, 9-Hexachlorodibenzofuran
CAS No: 92341-07-6
Formula: $C_{12}H_2Cl_6O$ MW: 372

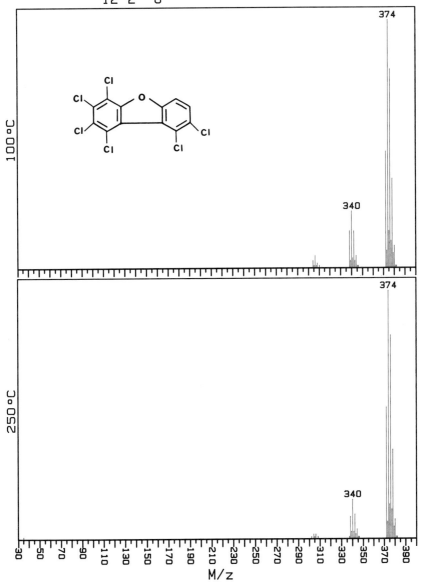

1, 2, 3, 4, 6, 7, 8-Heptachlorodibenzofuran
CAS No: 67562-39-4
Formula: $C_{12}HCl_7O$ MW: 406

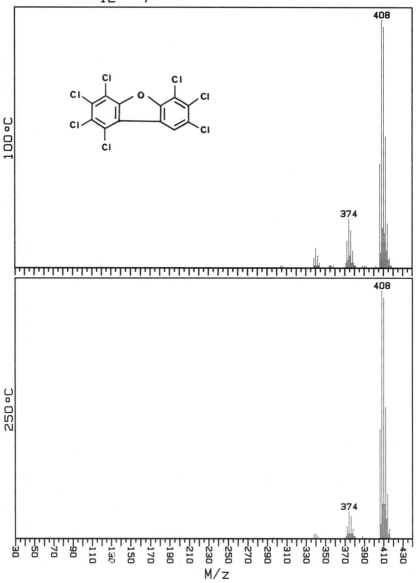

Octachlorodibenzofuran
CAS No: 39001-02-0
Formula: $C_{12}Cl_8O$ MW: 440

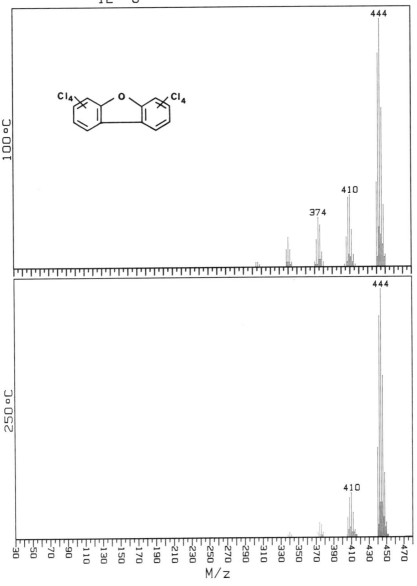

4-Chloroacetophenone
CAS No: 99-91-2
Formula: C_8H_7ClO MW: 154

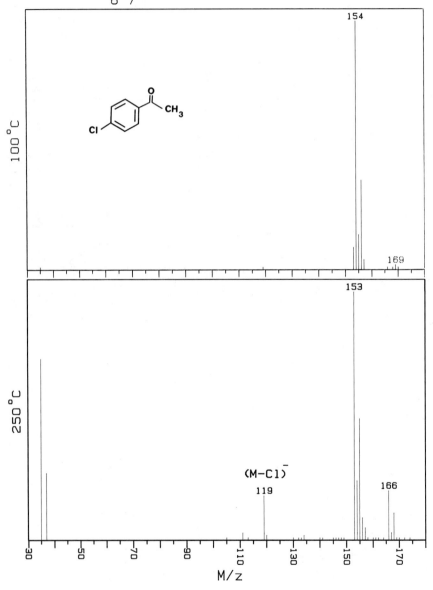

2-Chloroacetophenone
CAS No: 2142-68-9
Formula: C_8H_7ClO MW: 154

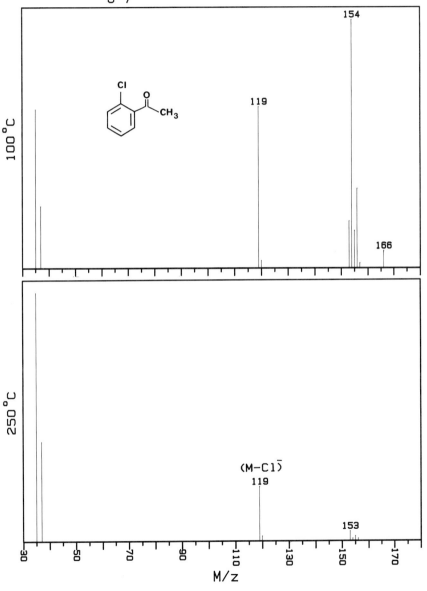

Benzophenone
CAS No: 119-61-9
Formula: $C_{13}H_{10}O$ MW: 182

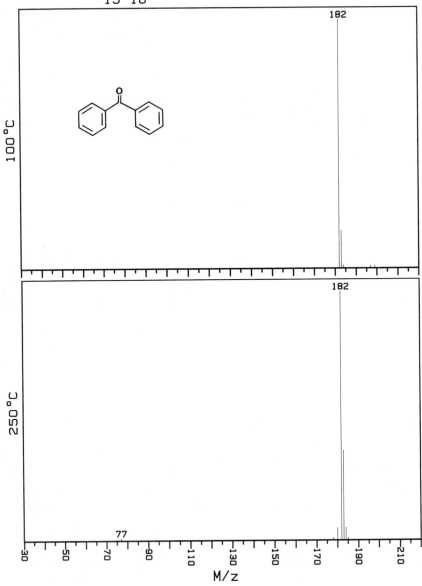

4-Chlorobenzophenone
CAS No: 134-85-0
Formula: $C_{13}H_9ClO$ MW: 216

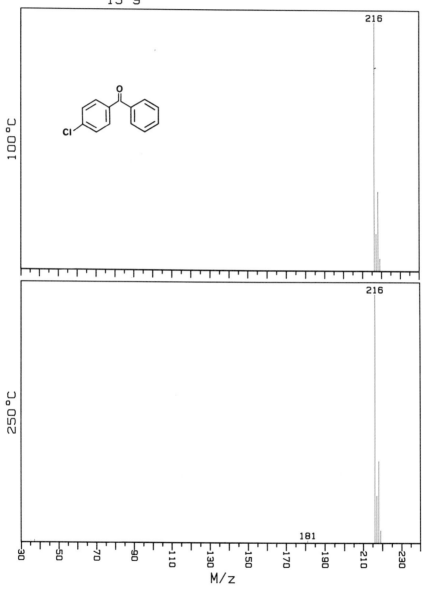

2-Chlorobenzophenone
CAS No: 5162-03-8
Formula: $C_{13}H_9ClO$ MW: 216

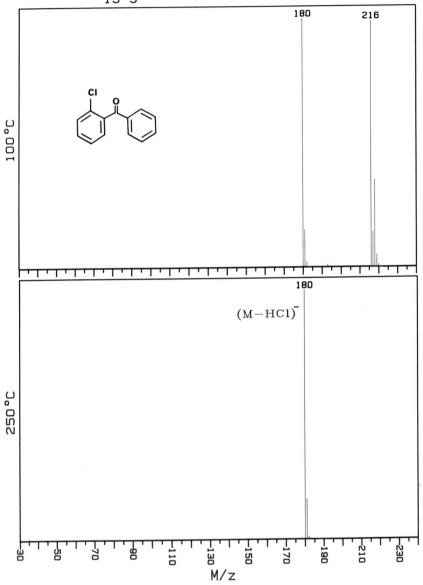

4,4'-Dichlorobenzophenone
CAS No: 90-98-2
Formula: $C_{13}H_8Cl_2O$ MW: 250

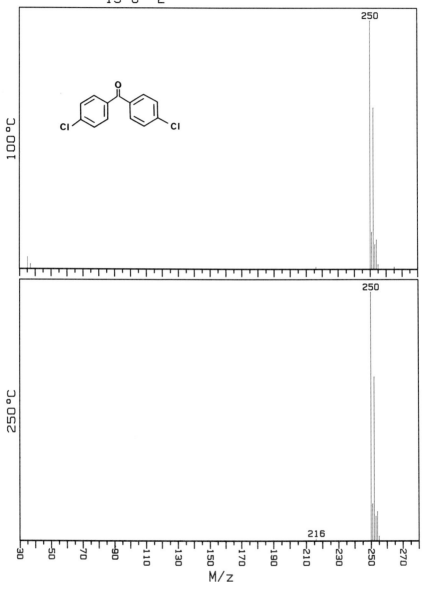

4,4'-Difluorobenzophenone
CAS No: 345-92-6
Formula: $C_{13}H_8F_2O$ MW: 218

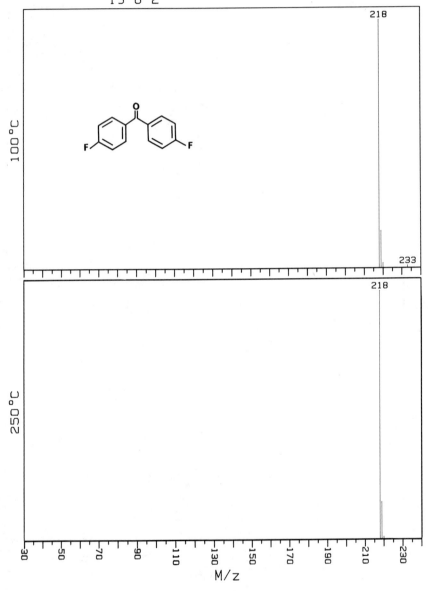

2, 3, 4, 5, 6-Pentafluorobenzophenone
CAS No: 1536-23-8
Formula: $C_{13}H_5F_5O$ MW: 272

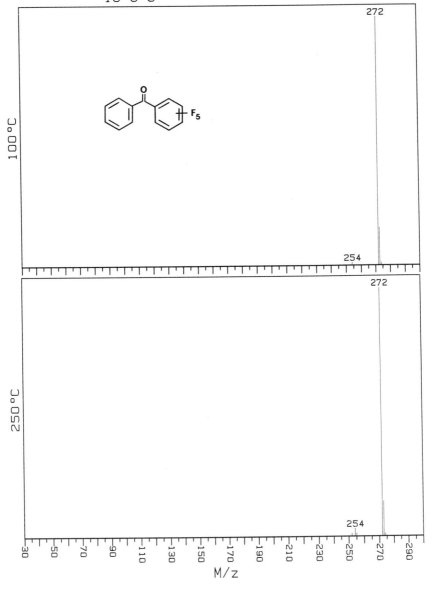

DDE-p, p′
CAS No: 72-55-9
Formula: $C_{14}H_8Cl_4$ MW: 316

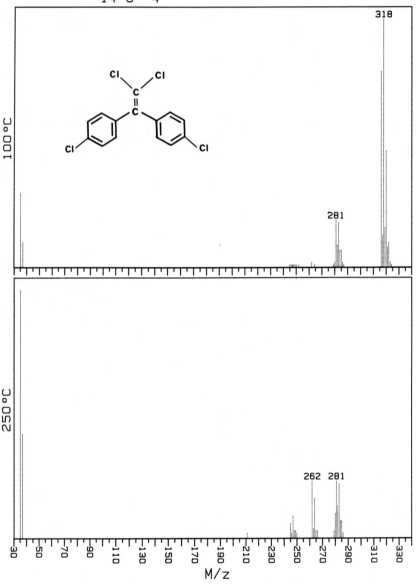

DDE-o,p'
CAS No: 3424-82-6
Formula: $C_{14}H_8Cl_4$ MW: 316

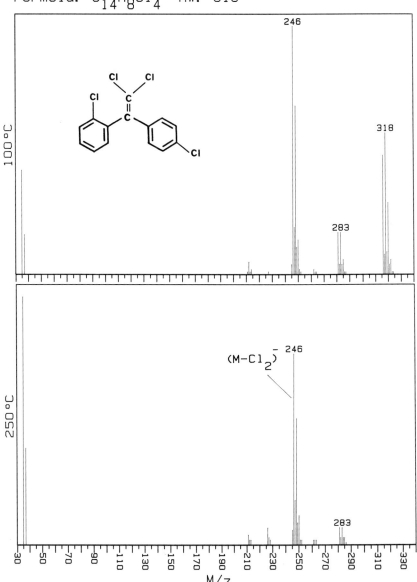

DDD-p, p'olefin
CAS No: 1022-22-6
Formula: $C_{14}H_9Cl_3$ MW: 282

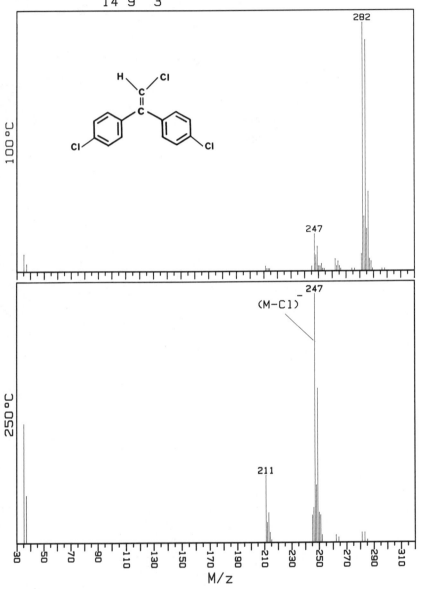

1, 1-Dichloro-2, 2-di-(4-tolyl)-ethylene
CAS No: 5432-01-9
Formula: $C_{16}H_{14}Cl_2$ MW: 276

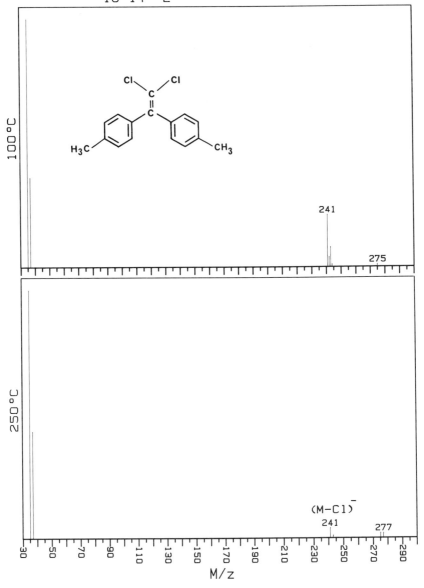

100 °C

241

275

250 °C

(M-Cl)$^-$
241

277

M/z

2,2-Bis-(4-fluorophenyl)-1,1,1-trifluoroethane
CAS No: 789-03-7
Formula: $C_{14}H_9F_5$ MW: 272

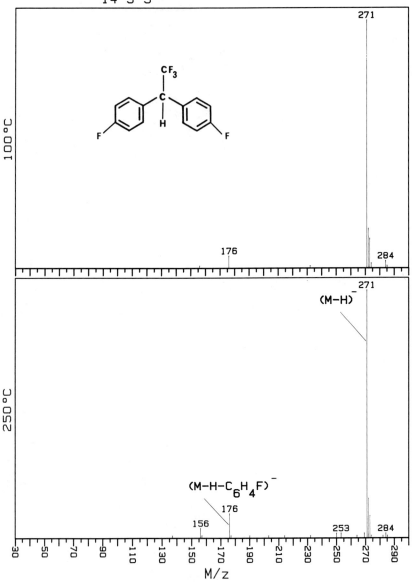

2, 2-Bis-(4-chlorophenyl)-1, 1, 1-trifluoroethane
CAS No: 361-07-9
Formula: $C_{14}H_9Cl_2F_3$ MW: 304

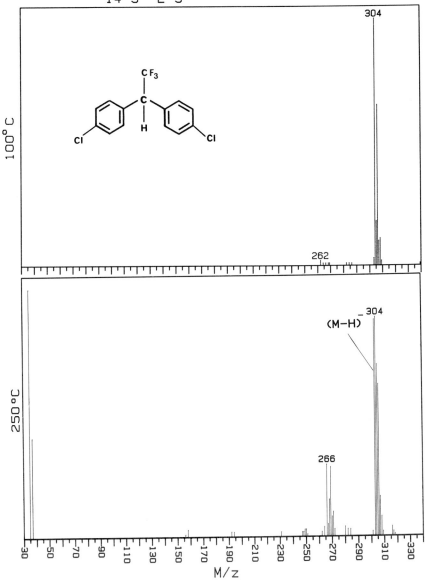

DDT-p, p′
CAS No: 50-29-3
Formula: $C_{14}H_9Cl_5$ MW: 352

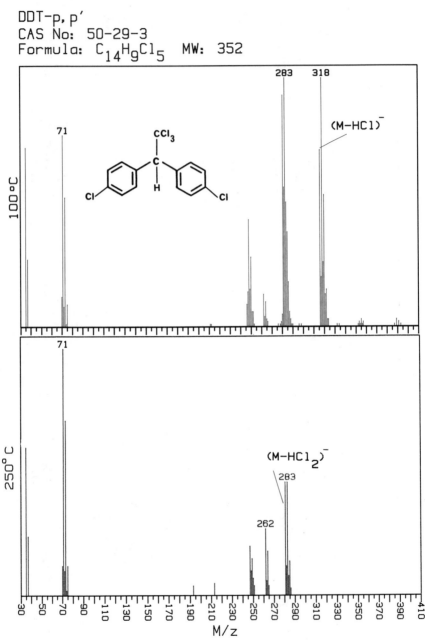

DDT-o,p'
CAS No: 789-02-6
Formula: $C_{14}H_9Cl_5$ MW: 352

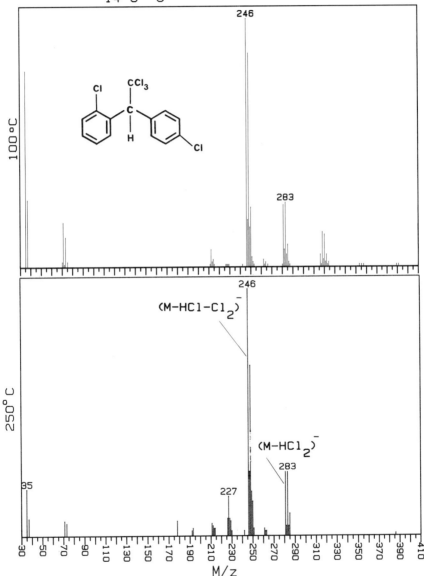

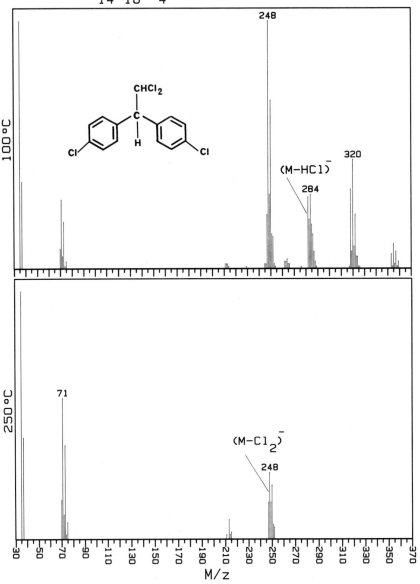

DDD-p, p′
CAS No: 72-54-8
Formula: $C_{14}H_{10}Cl_4$ MW: 318

DDD-o, p′
CAS No: 53-19-0
Formula: $C_{14}H_{10}Cl_4$ MW: 318

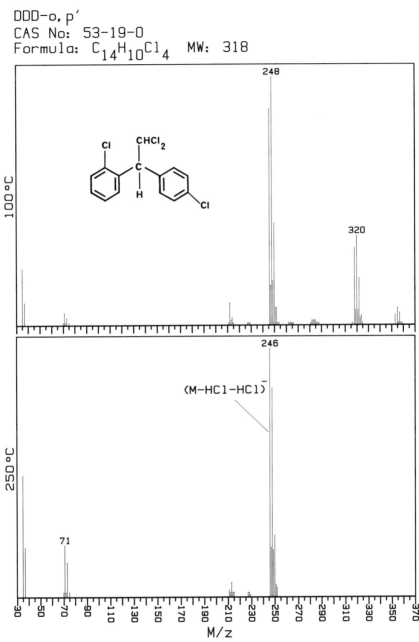

1,1-Bis-(4-chlorophenyl)-1,2,2,2-tetrachloroethane
CAS No: 3563-45-9
Formula: $C_{14}H_8Cl_6$ MW: 386

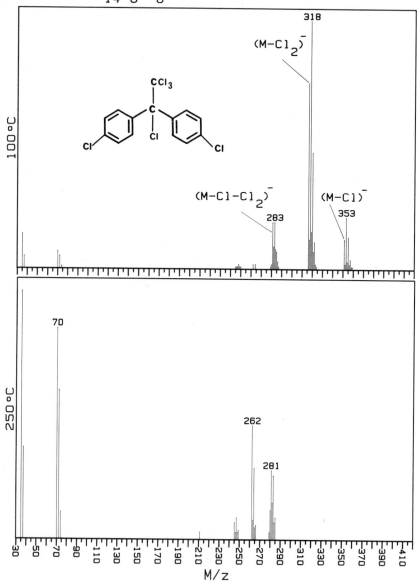

1-(4-Bromophenyl)-1-phenyl-2,2,2-trichloroethane
CAS No: 39211-93-1
Formula: $C_{14}H_{10}BrCl_3$ MW: 362

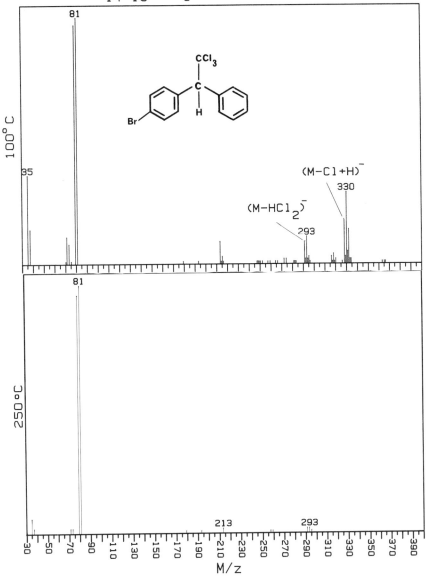

4,4'-Dichloro-a-(trifluoramethyl)benzhydrol
CAS No: 630-71-7
Formula: $C_{14}H_9Cl_2F_3O$ MW: 320

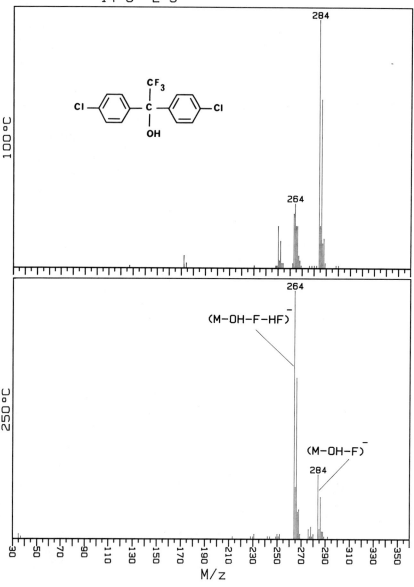

Dicofol
CAS No: 115-32-2
Formula: $C_{14}H_9Cl_5O$ MW: 368

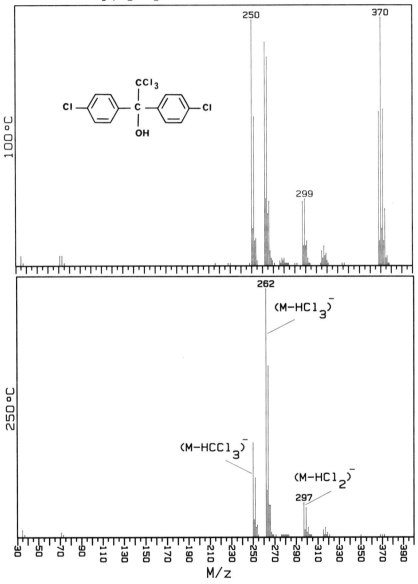

Chlorobenzilate
CAS No: 510-15-6
Formula: $C_{16}H_{14}Cl_2O_3$ MW: 324

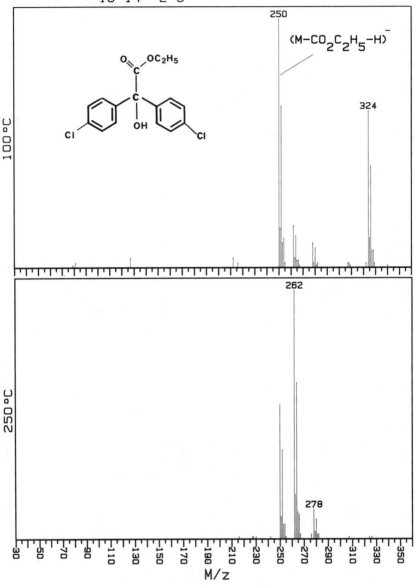

Bromopropylate
CAS No: 18181-80-1
Formula: $C_{17}H_{16}Br_2O_3$ MW: 426

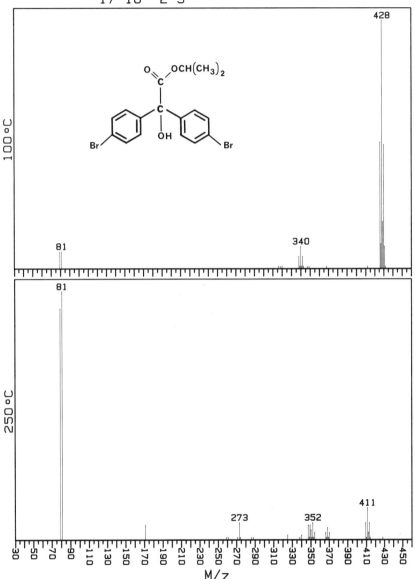

1-(4-Chlorophenyl)-2,2,2-trifluoroethanol
CAS No: 446-66-2
Formula: $C_8H_6ClF_3O$ MW: 210

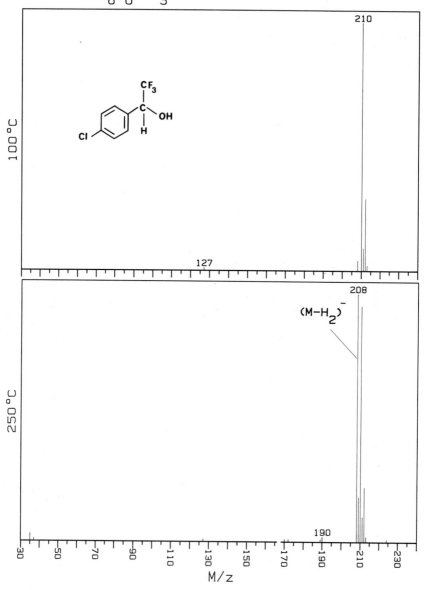

1-(4-Bromophenyl)-2,2,2-trichloroethanol
CAS No: 21757-86-8
Formula: $C_8H_6BrCl_3O$ MW: 302

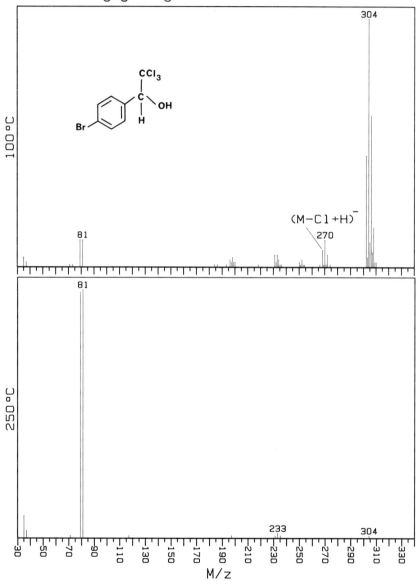

4, 4'-Dichlorobenzhydrol
CAS No: 90-97-1
Formula: $C_{13}H_{10}Cl_2O$ MW: 252

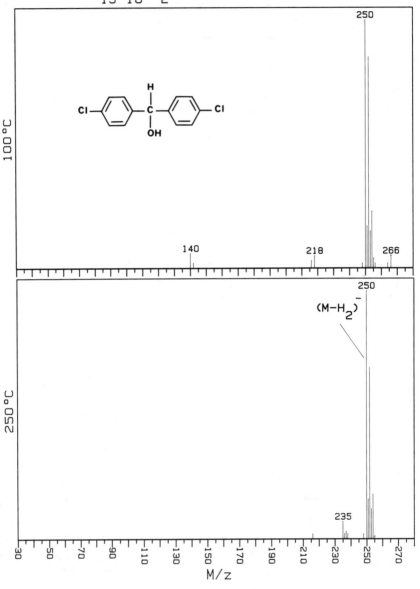

4.4′-Difluorobenzhydrol
CAS No: 365-24-2
Formula: $C_{13}H_{10}F_2O$ MW: 220

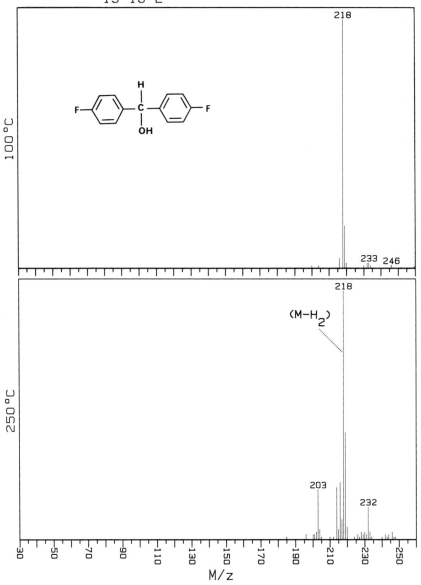

Aldrin
CAS No: 309-00-2
Formula: $C_{12}H_8Cl_6$ MW: 362

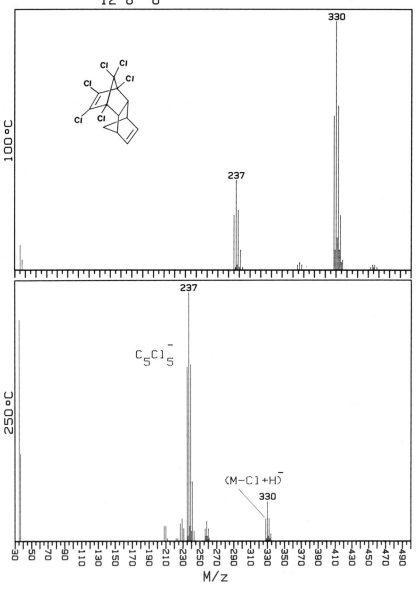

Isodrin
CAS No: 465-73-6
Formula: $C_{12}Cl_6H_8$ MW: 362

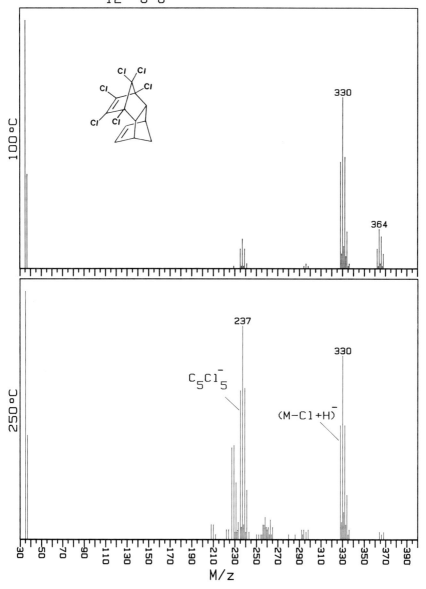

Dieldrin
CAS No: 60-57-1
Formula: $C_{12}H_8Cl_6O$ MW: 378

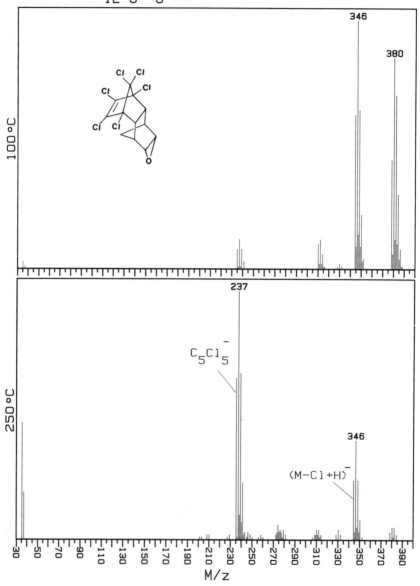

Endrin
CAS No: 72-20-8
Formula: $C_{12}H_8Cl_6O$ MW: 378

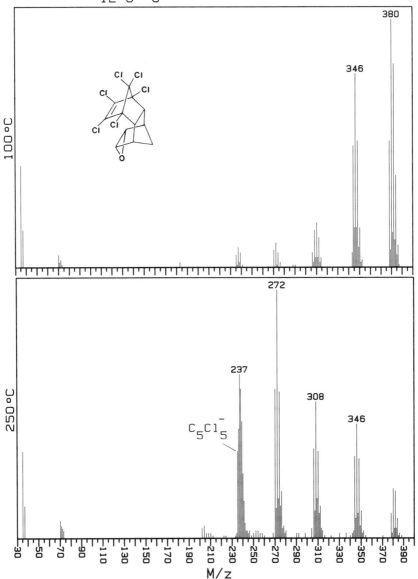

Endrin Aldehyde
CAS No: 7421-93-4
Formula: $C_{12}H_8Cl_6O$ MW: 378

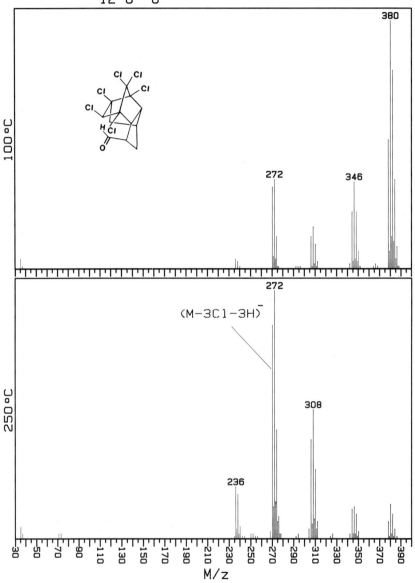

Endrin Ketone
CAS No: 53494-70-5
Formula: $C_{12}H_8Cl_6O$ MW: 378

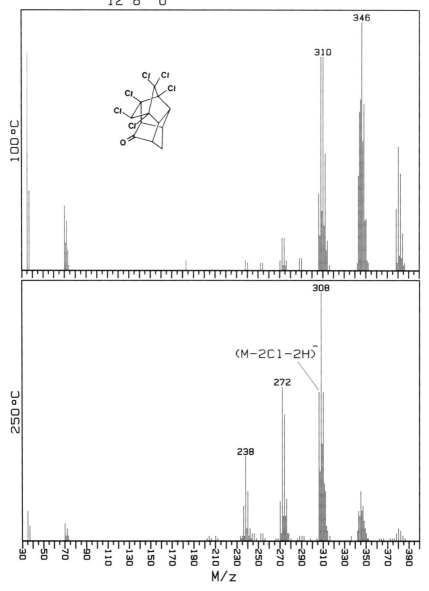

Chlordene
CAS No: 3734-48-3
Formula: $C_{10}H_6Cl_6$ MW: 336

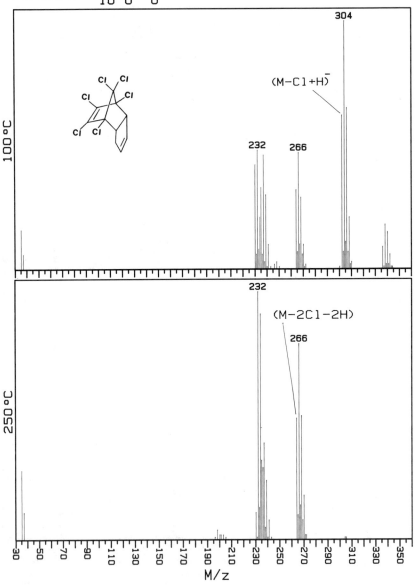

Chlordene, alpha
CAS No: 56534-02-2
Formula: $C_{10}H_6Cl_6$ MW: 336

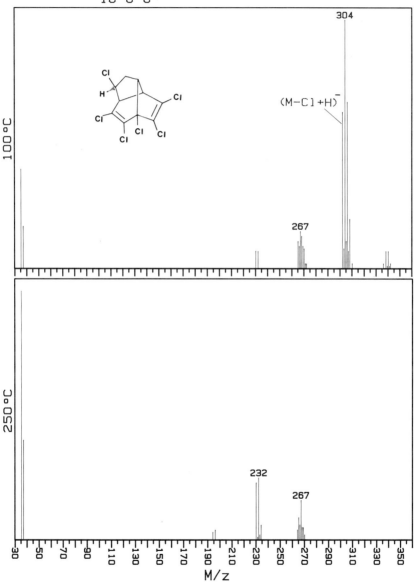

Chlordene, gamma
CAS No: 56641-38-4
Formula: $C_{10}C_6H_6$ MW: 336

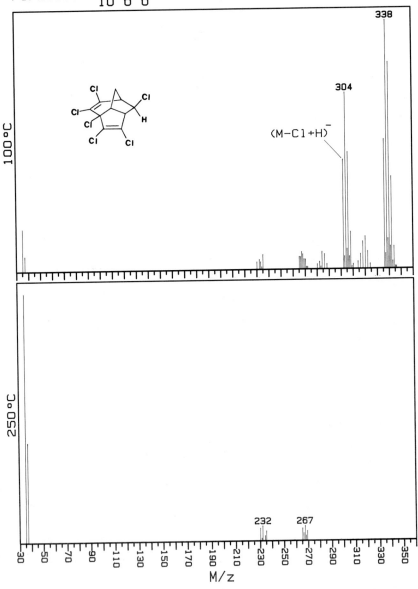

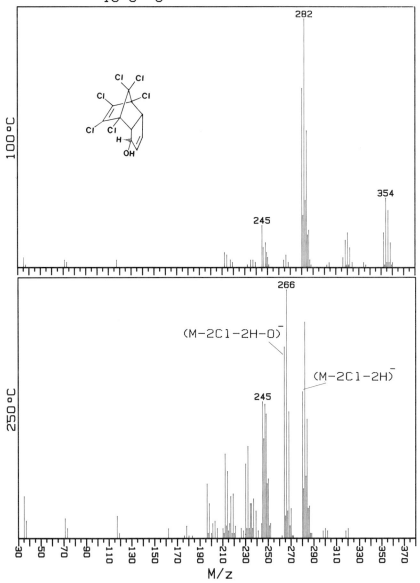

1-Hydroxychlordene
CAS No: 24009-05-0
Formula: $C_{10}H_6Cl_6O$ MW: 352

Heptachlor
CAS No: 76-44-8
Formula: $C_{10}H_5Cl_7$ MW: 370

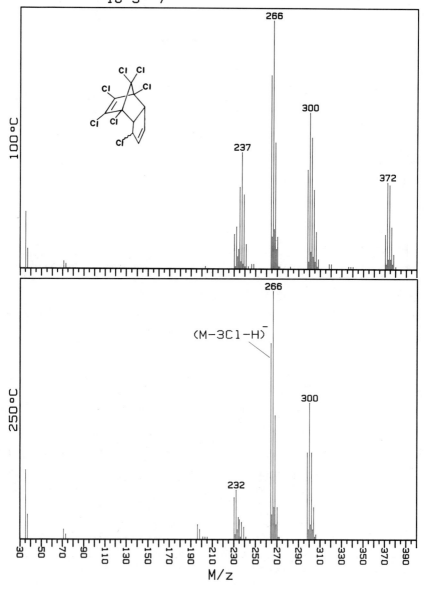

Heptachlor epoxide
CAS No: 1024-57-3
Formula: $C_{10}H_5Cl_7O$ MW: 386

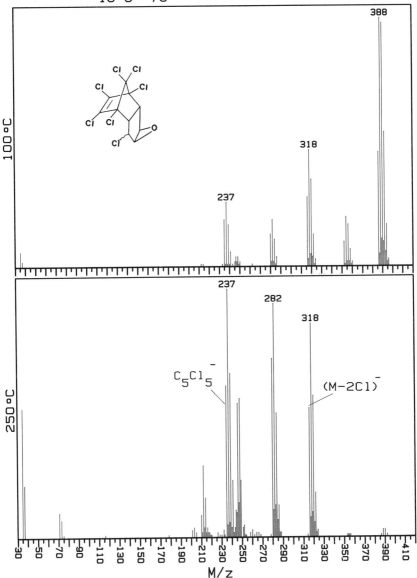

Oxychlordane
CAS No: 27304-13-8
Formula: $C_{10}H_4Cl_8O$, MW: 420

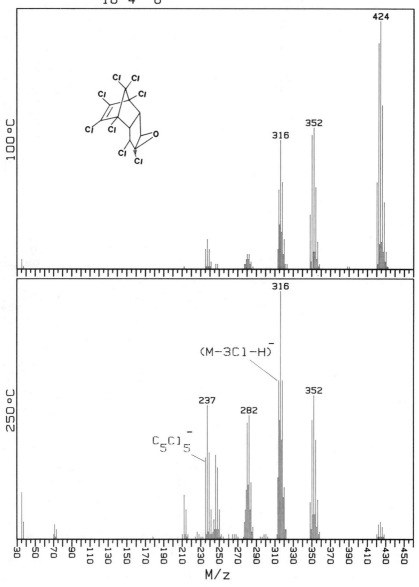

M/z

Chlordane, alpha
CAS No: 5103-71-9
Formula: $C_{10}H_6Cl_8$ MW: 406

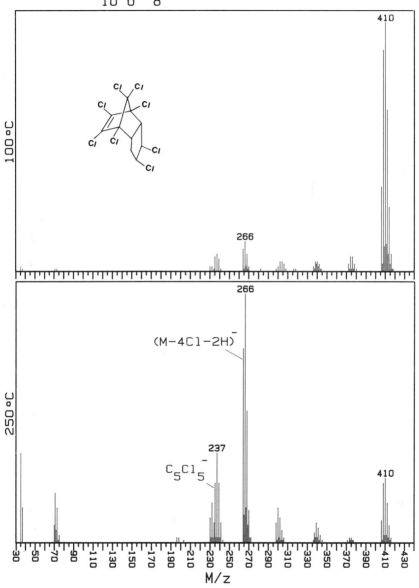

Chlordane, gamma
CAS No: 5566-34-7
Formula: $C_{10}H_6Cl_8$ MW: 406

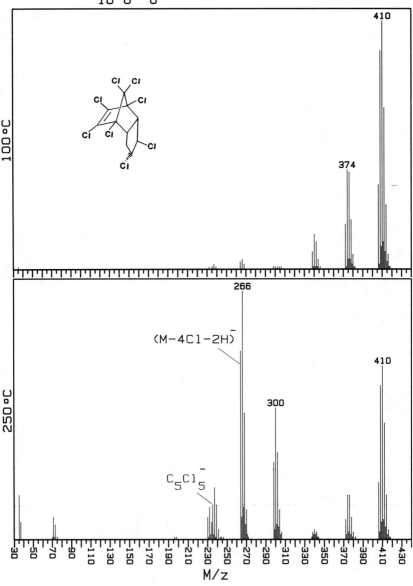

410

100 °C

374

266

(M-4Cl-2H)$^-$

300

410

250 °C

$C_5Cl_5^-$

M/z

trans-Nonachlor
CAS No: 39765-80-5
Formula: $C_{10}H_5Cl_9$, MW: 440

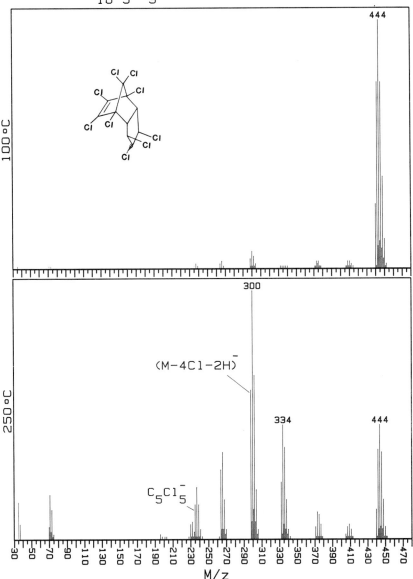

Pentac
CAS No: 2227-17-0
Formula: $C_{10}Cl_{10}$, MW: 470

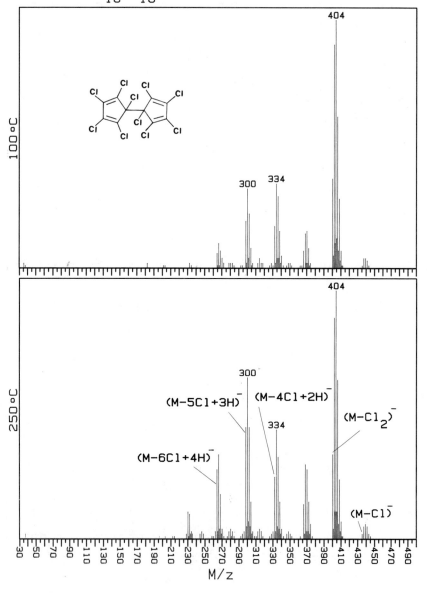

Mirex
CAS No: 2385-85-5
Formula: $C_{10}Cl_{12}$ MW: 540

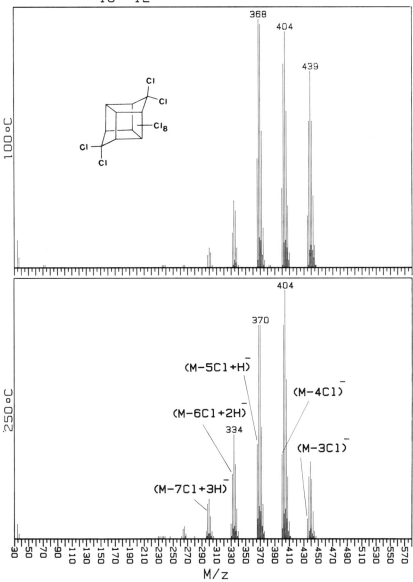

Photomirex
CAS No: 39801-14-4
Formula: $C_{10}H_1Cl_{11}$ MW: 506

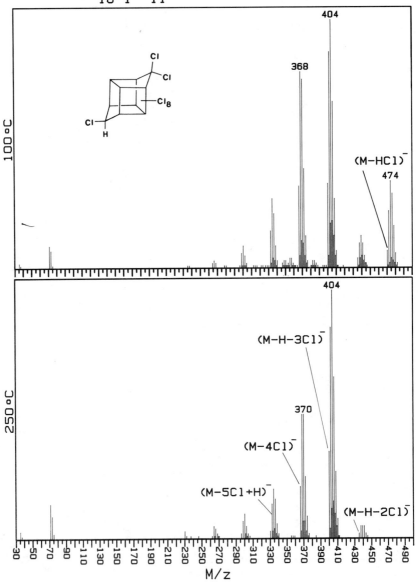

Kepone
CAS No: 143-50-0
Formula: $C_{10}Cl_{10}O$ MW: 486

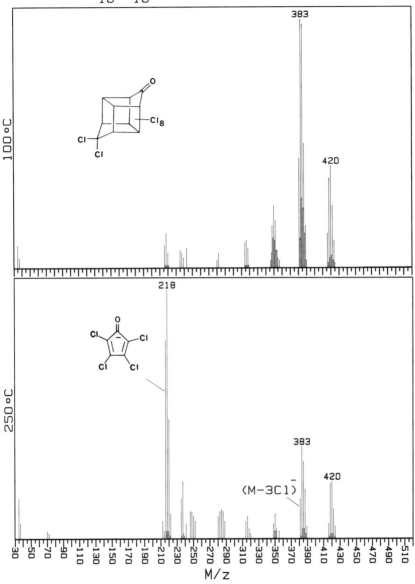

1,4,5,6,7,7-Hexachloro-5-norbornene-2,3-dicarboxylic anhydride
CAS No: 115-27-5
Formula: $C_9H_2Cl_6O_3$ MW: 368

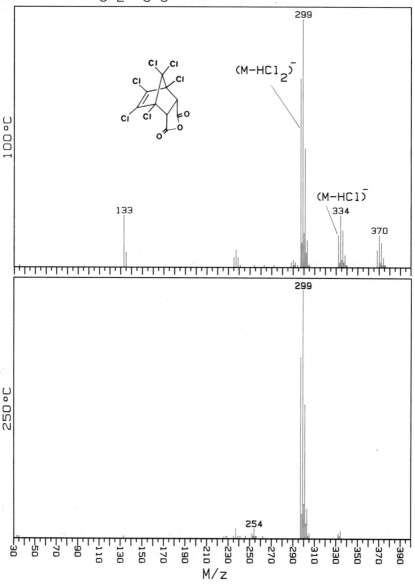

1,4,5,6,7,7-Hexachlorobicyclo[2.2.1]-5-heptene-2,3-dicarboximide
CAS No: 6889-41-4
Formula: $C_9H_3Cl_6NO_2$ MW: 367

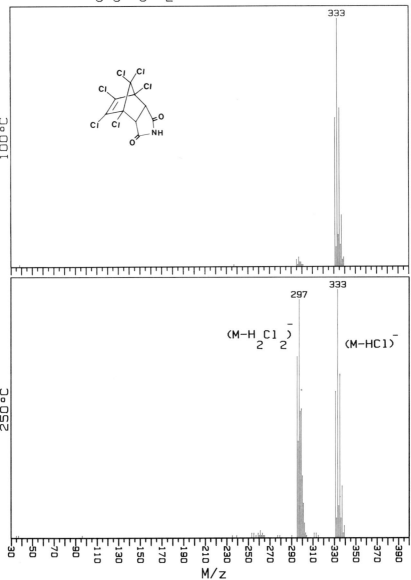

Dibutyl Chlorendate
CAS No: 1770-80-5
Formula: $C_{17}H_{20}Cl_6O_4$ MW: 498

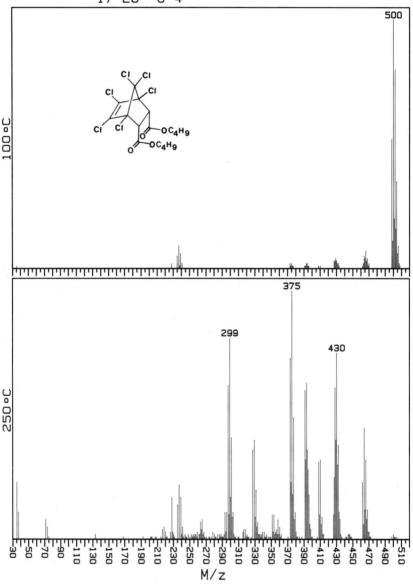

Chlorbicyclen
CAS No: 2550-75-6
Formula: $C_9H_6Cl_8$ MW: 394

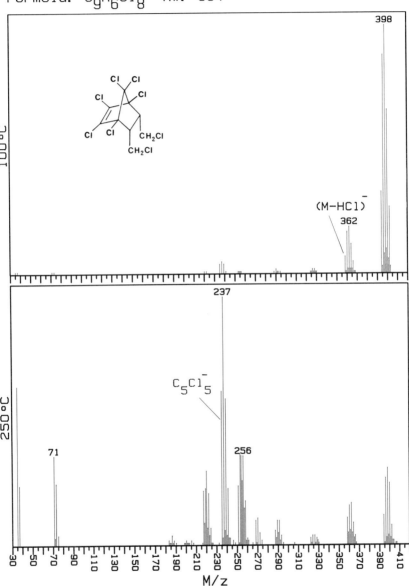

Endosulfan I
CAS No: 959-98-8
Formula: $C_9H_6Cl_6O_3S$ MW: 404

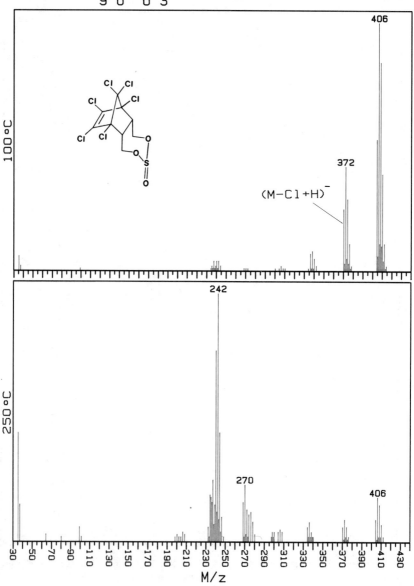

Endosulfan II
CAS No: 33213-65-9
Formula: $C_9H_6Cl_6O_3S$ MW: 404

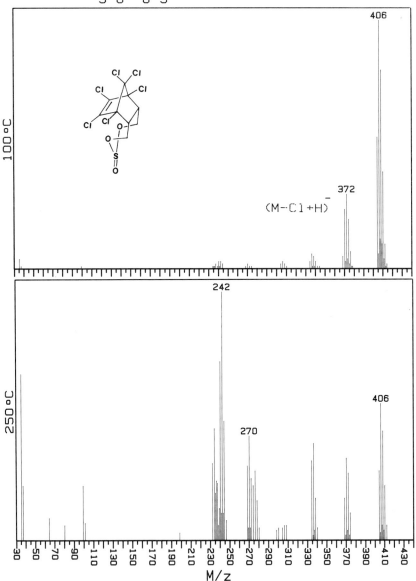

Endosulfan cyclic sulfate
CAS No: 1031-07-8
Formula: $C_9H_6Cl_6O_4S$ MW: 420

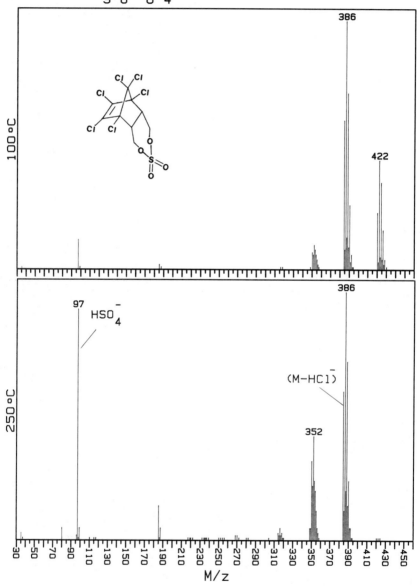

Tetradifon
CAS No: 116-29-0
Formula: $C_{12}H_6Cl_4O_2S$ MW: 354

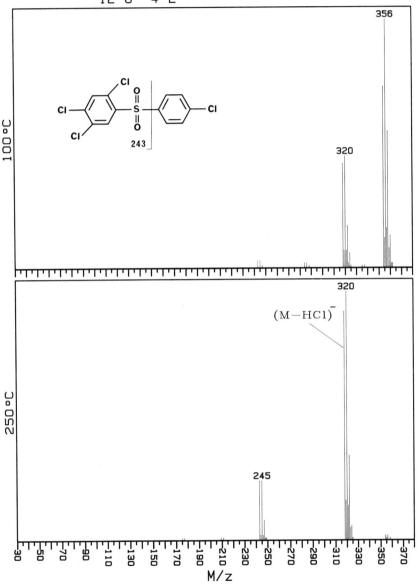

2-Chloroethyl-p-toluenesulfonate
CAS No: 80-41-1
Formula: $C_9H_{11}ClO_3S$ MW: 234

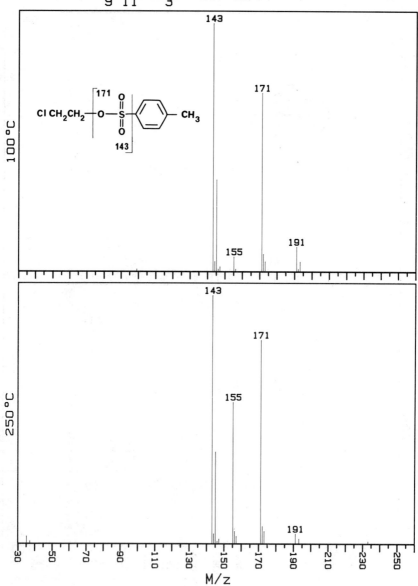

4,4′-Dichlorochalcone
CAS No: 19672-59-4
Formula: $C_{15}H_{10}Cl_2O$ MW: 276

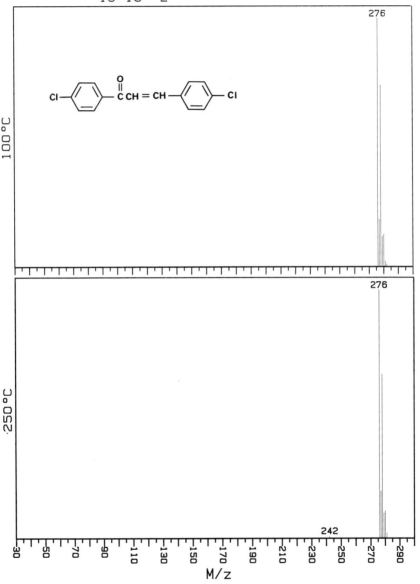

3, 3', 4, 4'-Tetrachloroazobenzene
CAS No: 14047-09-7
Formula: $C_{12}H_6Cl_4N_2$ MW: 318

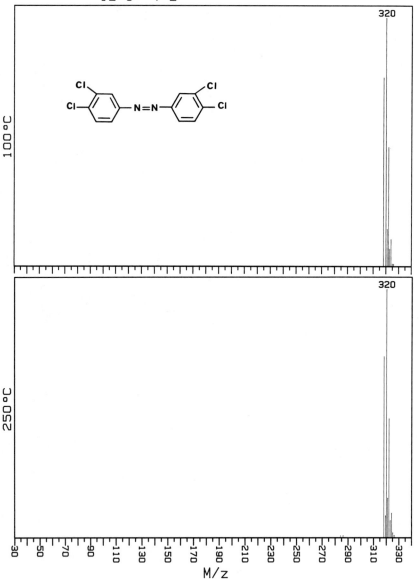

Decafluorotriphenylphosphine
CAS No: 5074-71-5
Formula: $C_{18}H_5F_{10}P$ MW: 442

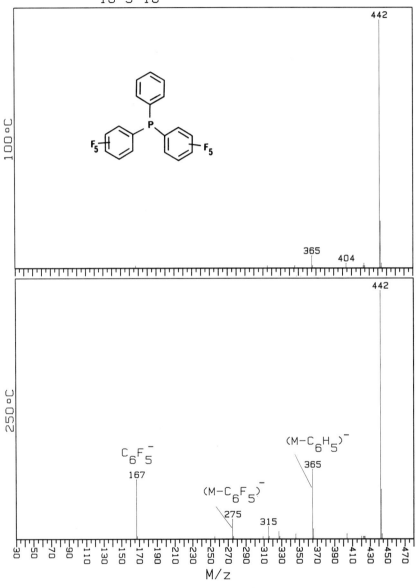

Decafluorotriphenylphosphine oxide
CAS No: not aval.
Formula: $C_{18}H_5F_{10}OP$ MW: 458

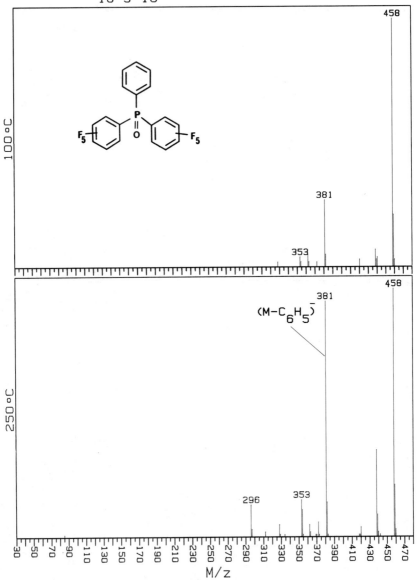

Perfluorotributylamine
CAS No: 311-89-7
Formula: $C_{12}F_{27}N$ MW: 671

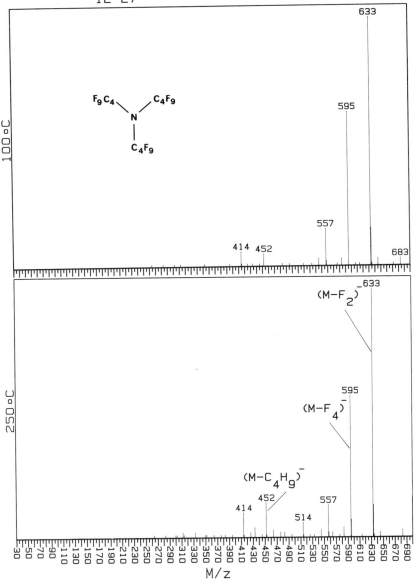

Perfluorokerosene-L
CAS No: not aval.
Formula: $CF_3(CF_2)_nCF_3$

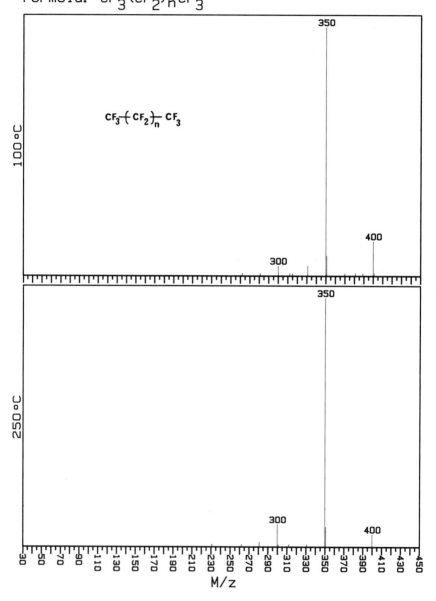

Perfluorokerosene-H
CAS No: not avail
Formula: $CF_3(CF_2)_nCF_3$

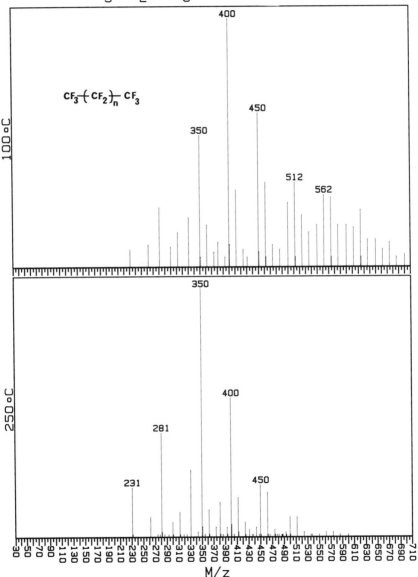

COMPOUND NAME INDEX

COMPOUND NAME INDEX

COMPOUND NAME INDEX

COMPOUND NAME INDEX

COMPOUND NAME INDEX

COMPOUND NAME INDEX

COMPOUND NAME INDEX

MOLECULAR WEIGHT INDEX

MOLECULAR WEIGHT INDEX

MOLECULAR WEIGHT INDEX

MOLECULAR WEIGHT INDEX

MOLECULAR WEIGHT INDEX

MOLECULAR WEIGHT INDEX

MOLECULAR WEIGHT INDEX

MW	FORMULA	NAME	PAGE
378	C12H8Cl6O	Endrin Ketone	331
386	C14H8Cl6	1,1-Bis(4-chlorophenyl)-1,2,2,2-tetrachloroethane	316
386	C10H5Cl7O	Heptachlor epoxide	337
388	C12H2Cl6O2	1,2,3,4,7,8-Hexachlorodibenzo-p-dioxin	290
388	C12H7Br3	2,4,5-Tribromobiphenyl	256
390	C6H2Br4	1,2,4,5-Tetrabromobenzene	43
390	C24H38O4	Di-n-octyl-phthalate	217
390	C24H38O4	Dioctyl phthalate	216
391	C19H29N5O4	4-Cyano-2,6-dinitro-N,N,N',N'-tetrapropyl-1,3-benzenediamine	204
392	C12H3Cl7	2,2',3,4,4',5,5'-Heptachlorobiphenyl	250
394	C9H6Cl8	Chlorbicyclen	349
400	C10Cl8	Octachloronaphthalene	238
404	C9H6Cl6O3S	Endosulfan I	350
404	C9H6Cl6O3S	Endosulfan II	351
406	C12HCl7O	1,2,3,4,6,7,8-Heptachlorodibenzofuran	296
406	C10H6Cl8	alpha-Chlordane	339
406	C10H6Cl8	gamma-Chlordane	340
409	C19H31N5O5	4-Carbamyl-2,6-dinitro-N,N,N',N'-tetrapropyl-1,3-benzenediamine	205
418	C26H42O4	Dinonyl phthalate	218
420	C9H6Cl6O4S	Endosulfan cyclic sulfate	352
420	C10H4Cl8O	Oxychlordane	338
422	C12HCl7O2	1,2,3,4,6,7,8-Heptachlorodibenzo-p-dioxin	291
426	C12H2Cl8	2,2',3,3',4,4',5,5'-Octachlorobiphenyl	251
426	C17H16Br2O3	Bromopropylate	321
440	C12Cl8O	Octachlorodibenzofuran	297
440	C10H5Cl9	trans-Nonachlor	341
442	C18H5F10P	Decafluorotriphenylphosphine	357
456	C12Cl8O2	Octachlorodibenzo-p-dioxin	292
458	C18H5F10OP	Decafluorotriphenylphosphine oxide	358
460	C12HCl9	2,2',3,3',4,4',5,5',6-Nonachloro-biphenyl	252
466	C12H6Br4	2,2',5,5'-Tetrabromobiphenyl	257
470	C10Cl10	Pentac	342
482	C7H3Br5	Pentabromotoluene	96
482	C12H6Br4O	Tetrabromodiphenyl ether (Bromkal 50-5 DE)	278
484	C6HBr5O	Pentabromophenol	71
486	C10Cl10O	Kepone	345
494	C12Cl10	Decachlorobiphenyl	253
498	C17H20Cl6O4	Dibutyl chlorendate	348
506	C10HCl11	Photomirex (2,8-dihydro mirex)	344
510	C12Cl10O	Decachlorodiphenyl ether	276
532	C2I4	Tetraiodoethene	13
540	C10Cl12	Mirex	343
544	C12H5Br5	2,2',4,5,5'-Pentabromobiphenyl	258
546	C6Br6	Hexabromobenzene	44
560	C12H5Br5O	Pentabromodiphenyl ether (Bromkal 70-5 DE)	279
622	C12H4Br6	2,2',4,4',5,5'-Hexabromobiphenyl	259
638	C12H4Br6O	Hexabromodiphenyl ether (Bromkal 70-5 DE)	280
671	C12F27N	Perfluorotributylamine	359

SPECTRAL PEAK INDEX AT 100 DEG.

m/z	int	m/z	int	m/z	int	m/z	int	ABBREVIATED NAME	PAGE
35	100	37	36	197	12	195	11	a,a-Dichlorotoluene	87
35	100	37	36	197	15	195	14	a,4-Dichlorotoluene	88
35	100	37	36	241	21	243	8	1,1-Dichloro-2,2-di-(4-tolyl)-	309
35	100	37	36	262	2	264	1	Hexachloro-1,3-butadiene	16
35	100	37	38	70	6	72	5	1,1,1,2-Tetrachloroethane	3
35	100	37	38	159	7	161	3	a,2-Dichlorotoluene	89
35	100	37	39	134	5	242	4	a,a,a,a',a',a'-Hexachloro-m-xylene	100
35	100	37	40	61	1	0	0	1,1,1-Trichloroethane	1
35	100	37	40	70	29	71	22	Pentachloroethane	5
35	100	37	40	124	5	174	4	a,2,6-Trichlorotoluene	92
35	100	37	41	0	0	0	0	a,a,a-Trichlorotoluene	90
35	100	37	43	70	2	122	1	Hexachloropropene	15
35	100	130	48	132	47	37	37	Trichloroethene	10
35	100	181	75	179	67	180	35	1,2,3-Trichlorobenzene	33
35	100	181	78	179	70	37	38	1,3,5-Trichlorobenzene	34
35	100	196	72	198	46	161	44	1,8-Dichloronaphthalene	232
35	100	237	62	239	39	235	37	Hexachlorocyclopentadiene	21
35	100	330	69	332	45	328	43	Isodrin	327
52	100	254	67	216	26	253	19	2,4,4'-Trichloro-2'-hydroxydiphenyl	281
59	100	138	8	122	7	123	5	2-Nitrophenol, acetyl derivative	105
70	100	72	70	35	34	37	14	Octachloropropane	9
71	100	35	74	73	64	201	39	2,2',4-Trichlorodiphenyl ether	270
79	100	81	99	314	90	316	84	1,2,4-Tribromobenzene	42
81	100	79	92	392	24	390	24	2,4,5-Tribromobiphenyl	256
81	100	79	93	160	85	162	43	Dibromochloropropane	8
81	100	79	95	208	1	206	1	1-Bromonaphthalene	230
81	100	79	97	35	36	330	29	1-(4-Bromophenyl)-1-phenyl-2,2,2-	317
81	100	79	97	391	45	389	43	2,2',5,5'-Tetrabromobiphenyl	257
81	100	79	98	155	16	157	15	Bromobenzene	26
81	100	79	99	312	6	231	6	4,4'-Dibromobiphenyl	255
81	100	79	99	314	73	316	72	1,3,5-Tribromobenzene	41
102	100	104	86	115	48	103	14	Benzonitrile	28
108	100	106	88	93	49	120	24	Phenol	29
111	100	113	41	112	10	35	8	Chlorobenzene	25
123	100	124	7	127	4	125	1	Nitrobenzene	24
127	100	0	0	0	0	0	0	Iodobenzene	27
131	100	133	98	70	91	72	64	1,1,2,2-Tetrachloroethane	4
137	100	138	9	136	2	121	1	4-Nitrotoluene	128
137	100	138	9	139	1	121	1	3-Nitrotoluene	118
137	100	138	10	139	1	121	1	2-Nitrotoluene	109
138	100	139	8	122	4	324	2	4-Nitroaniline	129
138	100	139	8	140	1	122	1	2-Nitroaniline	110
138	100	139	8	140	1	122	1	3-Nitroaniline	119
139	100	140	7	138	1	123	1	4-Nitrophenol	127
139	100	140	8	123	2	122	2	2-Nitrophenol	108
139	100	140	8	141	1	123	1	3-Nitrophenol	117
143	100	171	72	145	37	191	10	2-Chloroethyl-p-toluenesulfonate	354
145	100	147	62	35	22	149	12	1,3-Dichlorobenzene	31
145	100	147	64	35	20	149	9	1,2-Dichlorobenzene	30
145	100	147	67	149	12	35	11	1,4-Dichlorobenzene	32
148	100	149	9	132	4	130	2	4-Nitrobenzonitrile	121
148	100	149	9	132	5	117	2	2-Nitrobenzonitrile	102

SPECTRAL PEAK INDEX AT 100 DEG.

m/z	int	m/z	int	m/z	int	m/z	int	ABBREVIATED NAME	PAGE
148	100	149	9	132	6	116	2	3-Nitrobenzonitrile	112
151	100	152	9	135	3	153	1	4-Nitrobenzaldehyde	125
151	100	152	9	135	4	166	1	3-Nitrobenzaldehyde	116
151	100	152	9	153	1	135	1	2-Nitrobenzaldehyde	106
153	100	154	9	137	2	155	1	4-Methyl-2-nitrophenol	130
153	100	154	9	155	1	0	0	2-Nitroanisole	107
153	100	154	9	155	1	138	1	4-Nitroanisole	126
153	100	154	13	168	1	155	1	1-Cyanonaphthalene	228
154	100	119	65	35	64	156	32	2-Chloroacetophenone	299
154	100	156	36	155	14	153	9	4-Chloroacetophenone	298
156	100	157	14	153	2	171	1	1-Naphthaldehyde	229
157	100	159	33	158	7	160	2	1-Chloro-2-nitrobenzene	49
157	100	159	34	158	7	160	2	1-Chloro-4-nitrobenzene	51
157	100	159	34	158	8	160	3	1-Chloro-3-nitrobenzene	50
160	100	162	51	158	50	81	14	Tetrabromoethene	12
161	100	81	67	80	61	163	51	Tetrabromodiphenyl ether	278
161	100	81	95	79	89	405	54	Pentabromodiphenyl ether	279
163	100	161	73	165	26	162	17	1-Chloronaphthalene	231
163	100	195	19	182	10	164	8	Pentafluoroaniline	86
164	100	183	73	184	39	136	15	Pentafluoroanisole	79
164	100	184	32	165	7	146	3	Pentafluorophenol	72
164	100	194	89	148	45	179	38	Dimethyl phthalate	213
166	100	70	96	164	80	35	72	Hexachloroethane	6
166	100	164	81	168	50	170	11	Tetrachloroethene	11
167	100	168	11	154	6	166	4	2,4-Dinitrophenol,	141
167	100	184	44	154	35	168	25	2,4-Dinitrophenol	143
167	100	294	20	127	18	168	7	Iodopentafluorobenzene	48
168	100	138	23	152	9	169	8	1,3-Dinitrobenzene	111
168	100	169	8	138	6	152	2	1,2-Dinitrobenzene	101
168	100	169	8	138	6	152	4	1,4-Dinitrobenzene	120
169	100	167	74	171	53	35	53	1,1,2-Trichloroethane	2
173	100	174	12	157	2	175	1	1-Nitronaphthalene	227
174	100	176	76	35	70	193	54	a,a,2,6-Tetrachlorotoluene	93
176	100	178	80	205	59	207	40	Chloroneb	80
178	100	148	56	222	49	193	33	Diethyl phthalate	214
180	100	216	99	218	35	181	15	2-Chlorobenzophenone	302
181	100	179	99	35	87	195	43	1,2,4-Trichlorobenzene	35
181	100	182	9	165	8	149	4	4-Nitrobenzoic acid, methyl ester	123
181	100	182	9	183	1	0	0	3-Nitrophenol, acetyl derivative	115
181	100	182	10	138	4	183	1	4-Nitrophenol, acetyl derivative	124
181	100	182	11	165	4	183	1	3-Nitrobenzoic acid, methyl ester	114
181	100	182	11	183	1	138	1	2-Nitrobenzoic acid, methyl ester	104
181	100	182	78	163	43	162	38	Pentafluorotoluene	94
182	100	152	10	183	9	166	4	2,4-Dinitrotoluene	144
182	100	183	9	152	4	166	2	2,3-Dinitrotoluene	164
182	100	183	9	152	6	166	2	2,6-Dinitrotoluene	161
182	100	183	15	199	1	197	1	Benzophenone	300
183	100	153	17	184	8	167	4	2,6-Dinitroaniline	163
183	100	166	12	184	8	153	7	2,4-Dinitroaniline	149
183	100	181	73	185	50	35	16	1,2,3-Trichloropropane	7
183	100	184	8	153	7	167	3	3,5-Dinitroaniline	136
186	100	187	6	167	1	0	0	Hexafluorobenzene	45

378

SPECTRAL PEAK INDEX AT 100 DEG.

m/z	int	m/z	int	m/z	int	m/z	int	ABBREVIATED NAME	PAGE
187	100	221	59	223	54	35	46	2,4'-Dichlorobiphenyl	243
191	100	192	8	175	4	161	1	2-Nitrobenzotrifluoride	103
191	100	192	9	175	4	193	1	3-Nitrobenzotrifluoride	113
191	100	192	10	193	1	175	1	4-Nitrobenzotrifluoride	122
192	100	194	36	228	34	226	26	Dichlone	240
193	100	163	30	194	10	164	8	3,5-Dinitrobenzonitrile	131
193	100	194	8	175	1	0	0	Pentafluorobenzonitrile	60
193	100	195	98	35	65	197	37	2,4,5-Trichlorotoluene	91
194	100	195	13	196	1	0	0	Dimethyl terephthalate	221
196	100	198	99	200	33	199	8	2,3,5-Trichlorophenol	62
197	100	195	97	199	32	196	9	3,4,5-Trichloroaniline	81
197	100	198	9	167	9	199	1	2-Amino-4,6-dinitrotoluene	155
197	100	198	10	181	5	167	5	2,4-Dinitro-N-methylaniline	148
198	100	181	13	199	9	168	4	2,6-Dinitro-p-creosol	159
198	100	181	62	168	29	182	27	DNOC	184
198	100	183	70	168	13	199	9	2,4-Dinitroanisole	142
198	100	196	97	200	30	197	9	3,4,5-Trichlorophenol	66
198	100	196	97	200	30	197	7	2,4,5-Trichlorophenol	64
198	100	196	97	200	34	197	8	2,4,6-Trichlorophenol	65
198	100	196	99	200	33	199	7	2,3,6-Trichlorophenol	63
198	100	196	99	200	34	199	7	2,3,4-Trichlorophenol	61
198	100	199	9	182	4	168	3	3,5-Dinitrobenzyl alcohol	135
199	100	197	95	201	33	269	20	2,2,4,5-Tetrachlorocyclopentene-	23
202	100	203	20	216	2	204	2	Fluoranthene	224
202	100	204	32	203	7	205	2	Chloropentafluorobenzene	46
202	100	204	34	203	8	172	8	1-Chloro-2,4-dinitrobenzene	140
206	100	208	66	210	11	207	9	4,5-Dichloro-2-nitroaniline	150
206	100	208	67	210	11	207	9	2,4-Dichloro-6-nitroaniline	151
206	100	208	68	210	11	207	9	2,6-Dichloro-4-nitroaniline	152
210	100	212	29	211	9	208	4	1-(4-Chlorophenyl)-2,2,2-trifluoro-	322
211	100	181	8	195	3	179	2	2,6-Dinitro-N-ethylaniline	162
211	100	181	11	212	10	195	4	3,5-Dinitrobenzamide	133
212	100	210	99	214	34	211	10	2,4,6-Trichloroanisole	77
212	100	214	64	213	12	216	11	2,4-Dichloronapthol	241
215	100	213	74	231	48	217	48	2,3,5,6-Tetrachloronitrobenzene	56
215	100	217	67	219	11	201	11	2,4-Dichloro-3,5-dinitrobenzonitrile	165
216	100	214	72	218	46	215	14	1,2,4,5-Tetrachlorobenzene	36
216	100	214	77	218	48	220	10	1,2,3,4-Tetrachlorobenzene	37
216	100	218	32	217	15	219	5	4-Chlorobenzophenone	301
217	100	219	33	200	16	218	8	6-Chloro-2,4-dinitroaniline	153
217	100	219	36	218	20	35	12	Diclofop-methyl	285
218	100	216	81	220	51	222	11	Hexachloro-3-cyclopentenone	22
218	100	219	15	220	2	233	1	4,4'-Difluorobenzophenone	304
218	100	219	17	216	4	233	2	4,4'-Difluorobenzhydrol	325
219	100	277	85	247	80	230	31	N-(2-Bromoethyl)-2,6-dinitro-4-	208
221	100	222	36	312	16	248	10	Butyl benzyl phthalate	220
221	100	223	84	222	22	225	20	2,2'-Dichlorobiphenyl	244
223	100	221	67	225	40	187	18	4,4'-Dichlorobiphenyl	242
223	100	224	14	222	9	210	7	Dinoseb Acetate	187
224	100	226	98	228	31	240	26	1,3-Dinitro-2,4,5-trichlorobenzene	168
225	100	227	33	226	9	228	3	4-Chloro-3-nitrobenzotrifluoride	157
225	100	227	35	226	8	195	4	5-Chloro-2-nitrobenzotrifluoride	158

SPECTRAL PEAK INDEX AT 100 DEG.

m/z	int	m/z	int	m/z	int	m/z	int	ABBREVIATED NAME	PAGE
225	100	227	35	226	9	209	6	2-Chloro-5-nitrobenzotrifluoride	156
225	100	227	97	229	32	197	12	1,2,3-Trichloro-4-nitrobenzene	52
225	100	227	98	195	65	197	62	1,3,5-Trichloro-2-nitrobenzene	54
225	100	227	98	229	31	195	14	1,2,4-Trichloro-5-nitrobenzene	53
226	100	196	25	227	11	228	2	3,5-Dinitrobenzoic acid,	134
226	100	196	37	209	22	210	19	2,4-Dinitro-6-isopropylphenol	185
227	100	197	69	181	45	229	33	4-Chloro-3,5-dinitrobenzonitrile	137
230	100	231	19	247	2	232	2	Diphenylfulvene	223
230	100	232	99	234	32	231	13	Trichloronaphthalene (Halowax 1014)	233
231	100	229	76	233	50	235	11	2,3,4,5-Tetrachloroaniline	83
231	100	229	77	233	50	235	11	2,3,5,6-Tetrachloroaniline	84
231	100	230	96	232	93	229	82	Pentachlorophenol	70
232	100	230	74	234	52	236	11	2,3,5,6-Tetrachlorophenol	68
232	100	230	76	234	51	236	12	2,3,4,5-Tetrachlorophenol	67
232	100	234	66	70	50	72	32	2,3-Dichlorohexafluoro-2-butene	14
233	100	235	65	249	19	251	12	2,4-Dichloro-3,5-dinitrobenzamide	167
235	100	252	37	236	16	222	5	2,6-Dinitro-4-(trifluoromethyl)phenol	160
236	100	206	12	237	9	220	6	3,5-Dinitrobenzotrifluoride	132
239	100	240	13	209	6	241	2	N,N-Diethyl-2,4-dinitroaniline	146
240	100	210	39	208	20	224	17	Dinoseb	186
242	100	240	62	244	45	243	8	4-Bromo-2,5-dichlorophenol	74
242	100	240	77	244	53	243	48	a,a',2,3,5,6-Hexachloro-p-xylene	98
242	100	240	78	244	51	35	38	a,a,a,a',a',a'-Hexachloro-p-xylene	99
244	100	198	20	245	15	199	3	2,2'-Dinitrobiphenyl	260
244	100	242	78	246	49	243	13	2,4,5,6-Tetrachloro-m-xylene	97
245	100	247	32	179	20	246	11	N-(2-Chloroethyl)-2,4-dinitroaniline	145
246	100	248	68	318	57	316	48	DDE-o,p'	307
246	100	248	86	35	79	37	27	DDT-o,p'	313
246	100	248	99	247	8	249	7	Bromopentafluorobenzene	47
247	100	245	75	246	60	249	51	Pentachlorothiophenol	73
248	100	35	99	250	68	320	44	DDD-p,p'	314
248	100	248	87	250	41	320	36	DDD-o,p'	315
250	100	252	64	248	61	254	22	Pentachlorobenzene	38
250	100	252	65	251	15	254	12	4,4'-Dichlorobenzophenone	303
250	100	252	65	324	63	326	41	Chlorobenzilate	320
250	100	252	85	254	23	251	17	4,4'-Dichlorobenzhydrol	324
252	100	253	23	254	3	0	0	Benzo[a]pyrene	225
254	100	151	95	278	75	127	74	Tetraiodoethene	13
255	100	71	64	257	63	253	61	BHC, alpha isomer	17
255	100	71	89	257	66	253	60	BHC, beta isomer	18
255	100	257	64	253	61	259	22	BHC, delta isomer	19
255	100	257	68	71	58	253	57	BHC, gamma isomer	20
258	100	256	87	257	67	35	54	2,2',5-Trichlorobiphenyl	245
258	100	260	66	274	36	276	23	2,4-Dichloro-3,5-dinitrobenzo-	166
260	100	262	35	230	19	261	11	4-Chloro-3,5-dinitrobenzoic acid,	139
261	100	259	78	231	74	229	57	2,3,4,5-Tetrachloronitrobenzene	55
263	100	261	98	246	17	244	22	6-Bromo-2,4-dinitroaniline	154
264	100	266	64	262	61	268	21	Pentachlorotoluene	95
265	100	267	63	263	62	269	21	Pentachloroaniline	85
265	100	267	64	263	63	249	53	Pentachloronitrobenzene	57
265	100	479	75	267	64	263	63	Decachlorodiphenyl ether	276
266	100	264	75	268	47	267	12	Tetrachloronaphthalene (Halowax 1014)	234

SPECTRAL PEAK INDEX AT 100 DEG.

m/z	int	m/z	int	m/z	int	m/z	int	ABBREVIATED NAME	PAGE
266	100	264	77	268	46	270	11	Chlorothalonil	59
266	100	264	78	300	63	302	53	Heptachlor	336
267	100	268	14	237	6	269	2	2,4-Dinitro-N,N-dipropylaniline	147
270	100	204	64	272	34	234	16	N-(2-Chloroethyl)-4-cyano-2,6-	212
270	100	272	32	240	28	271	9	4-Chloro-3,5-dinitrobenzotrifluoride	138
270	100	272	94	274	32	271	17	2',3,4,4'-Tetrachlorodiphenyl ether	272
271	100	272	16	273	12	176	5	2,2-Bis(4-fluorophenyl)-1,1,1-	310
272	100	270	98	274	30	271	14	2,4,4',5-Tetrachlorodiphenyl ether	271
272	100	273	12	274	1	254	1	Octafluoronaphthalene	239
272	100	273	15	274	1	254	1	2,3,4,5,6-Pentafluorobenzophenone	305
272	100	273	15	274	2	256	1	2,2'-Dinitrobibenzyl	261
272	100	273	18	274	2	242	1	4,4'-Dinitrobibenzyl	262
275	100	277	68	273	60	279	22	Pentachlorobenzonitrile	58
276	100	277	25	278	3	252	1	Benzo[ghi]perylene	226
276	100	278	73	277	19	280	13	4,4'-Dichlorochalcone	355
277	100	198	77	196	73	275	50	Bromoxynil	75
278	100	206	86	148	68	279	25	Dibutyl phthalate	215
278	100	279	19	295	18	296	4	Dinocap 5	192
278	100	279	21	295	20	296	4	Dinocap 3	190
279	100	249	13	280	11	263	3	N,N-Dimethyl-2,6-dinitro-4-	173
279	100	249	45	280	14	250	6	4-Cyano-2,6-dinitro-N,N,N',N'-	199
279	100	278	70	295	60	280	19	Dinocap 4	191
279	100	280	11	249	7	263	5	2,6-Dinitro-N-ethyl-4-	180
279	100	295	50	280	19	278	13	Dinocap 1	188
279	100	295	51	280	18	278	15	Dinocap 2	189
280	100	282	66	278	65	284	22	Pentachloroanisole	78
281	100	282	16	251	9	249	4	Pendimethalin	182
282	100	280	72	284	55	354	28	1-Hydroxychlordene	335
282	100	284	93	286	32	283	22	DDD-p,p'olefin	308
283	100	285	61	284	14	287	12	Nitrofen	282
284	100	286	68	264	26	263	22	4,4'-Dichloro-a-(trifluoromethyl)-	318
284	100	286	82	282	51	288	37	Hexachlorobenzene	39
288	100	286	93	290	35	287	17	Tetrasul	277
288	100	286	99	324	90	322	70	3,3',5,5'-Tetrachloro-4,4'-biphenyldiol	265
292	100	290	81	294	52	293	24	2,2',5,5'-Tetrachlorobiphenyl	246
292	100	293	16	262	10	294	2	4-Cyano-2,6-dinitro-N,N-dipropyl-	179
293	100	294	13	263	6	277	5	2,6-Dinitro-N-propyl-4-	181
295	100	296	14	279	6	248	5	2,6-Dinitro-N-(2-hydroxyethyl)-4-	209
295	100	296	18	263	11	265	8	Butralin	183
297	100	237	30	298	15	221	10	4-Carbamyl-2,6-dinitro-N,N,N',N'-	200
299	100	297	76	301	48	334	21	1,4,5,6,7,7-Hexachloro-5-norbornene-	346
300	100	302	63	298	63	304	21	Pentachloronaphthalene (Halowax 1014)	235
303	100	305	35	304	13	237	12	N-(2-Chloroethyl)-2,6-dinitro-4-	211
304	100	305	15	306	5	138	5	1,2-Bis(4-nitrophenoxy)ethane	286
304	100	305	16	138	5	306	3	1,2-Bis(2-nitrophenoxy)ethane	287
304	100	306	61	302	45	308	16	1-(4-Bromophenyl)-2,2,2-trichloro-	323
304	100	306	65	302	62	232	48	Chlordene	332
304	100	306	65	305	18	308	11	2,2-Bis(4-chlorophenyl)-1,1,1-	311
304	100	306	67	302	63	35	40	alpha-Chlordene	333
306	100	304	71	308	46	307	14	1,2,7,8-Tetrachlorodibenzofuran	293
306	100	304	75	308	46	307	14	3,3',4,4',5-Pentachlorodiphenyl ether	275
306	100	304	78	308	50	307	17	2,2',4,4',5-Pentachlorodiphenyl ether	274

SPECTRAL PEAK INDEX AT 100 DEG.

m/z	int	m/z	int	m/z	int	m/z	int	ABBREVIATED NAME	PAGE
307	100	305	77	309	49	306	48	2',3,3',4',5'-Pentachloro-2-biphenyol	266
307	100	306	77	305	75	308	54	2',3',4',5,5'-Pentachloro-2-biphenyol	267
307	100	308	15	277	7	291	3	N,N-Diethyl-2,6-dinitro-4-	172
308	100	306	74	310	51	309	14	2',3',4',5'-Tetrachloro-3-biphenyol	264
308	100	306	76	310	51	309	14	2',3',4',5'-Tetrachloro-4-biphenyol	263
308	100	306	78	310	49	309	18	3,3',4,4'-Tetrachlorodiphenyl ether	273
309	100	310	19	311	3	279	3	Isopropalin	174
313	100	283	45	315	33	267	16	3-Chloro-N,N-dimethyl-2,6-dinitro-	195
313	100	315	34	247	24	219	19	N-(2-Chloroethyl)-2,6-dinitro-4-	207
318	100	148	33	319	22	320	3	Diphenyl phthalate	219
318	100	283	99	281	93	71	77	DDT-p,p'	312
318	100	316	75	320	47	353	21	1,1-Bis(4-chlorophenyl)-1,2,2,2-	316
318	100	316	79	320	47	35	30	DDE-p,p'	306
320	100	318	76	322	48	321	15	3,3',4,4'-Tetrachloroazobenzene	356
322	100	292	25	323	14	293	4	2,6-Dinitro-N,N,N',N'-tetramethyl-4-	198
322	100	320	74	324	50	323	14	1,2,3,4-Tetrachlorodibenzo-p-dioxin	288
322	100	323	15	292	7	232	7	Dinitramine	193
325	100	326	16	295	5	327	3	2,6-Dinitro-N,N-dipropyl-4-	178
326	100	328	66	324	62	330	22	2,2',4,5,5'-Pentachlorobiphenyl	247
326	100	328	67	324	66	330	22	2,3,4,5,6-Pentachlorobiphenyl	248
327	100	329	35	328	14	330	4	N-(3-Chloropropyl)-2,6-dinitro-4-	210
329	100	331	99	333	33	327	33	2,4,6-Tribromoaniline	82
330	100	332	66	328	62	237	36	Aldrin	326
330	100	332	97	328	35	334	33	2,4,6-Tribromophenol	69
332	100	330	72	334	47	333	12	Dacthal	222
333	100	335	64	331	60	337	21	1,4,5,6,7,7-Hexachlorobicyclo[2.2.1]-	347
334	100	332	99	336	33	333	22	2,4'',5-Trichloro-p-terphenyl	268
334	100	335	16	336	1	316	1	Decafluorobiphenyl	254
334	100	336	79	300	61	332	49	Hexachloronaphthalene (Halowax 1014)	236
335	100	336	17	305	5	319	3	Trifluralin	169
335	100	336	17	305	6	319	3	Benefin	170
335	100	336	19	305	10	337	3	4-Cyano-2,6-dinitro-N,N,N',N'-	202
338	100	340	83	304	71	336	52	gamma-Chlordene	334
340	100	342	67	338	60	344	21	1,2,3,8,9-Pentachlorodibenzofuran	294
341	100	311	49	343	33	313	16	3-Chloro-N,N-diethyl-2,6-dinitro-	196
341	100	343	67	342	16	345	11	Bifenox	284
345	100	346	17	347	7	315	4	Nitralin	177
346	100	35	88	310	86	308	86	Endrin Ketone	331
346	100	347	16	348	7	330	2	Oryzalin	175
346	100	380	85	382	70	348	64	Dieldrin	328
347	100	348	18	317	7	331	3	Profluralin	171
350	100	400	14	351	8	331	4	Perfluorokerosene-L	360
353	100	354	18	279	11	355	3	4-Carbamyl-2,6-dinitro-N,N,N',N'-	203
355	100	357	35	319	23	356	15	Fluchloralin	206
356	100	354	73	358	55	320	45	Tetradifon	353
356	100	358	67	354	59	360	22	1,2,3,4,7-Pentachlorodibenzo-p-dioxin	289
360	100	362	87	358	54	364	36	2,2',4,4',5,5'-Hexachlorobiphenyl	249
361	100	363	34	362	19	296	7	Oxyfluorfen	283
364	100	317	36	365	17	334	13	Dinitramine, acetyl derivative	194
368	100	366	76	370	50	369	20	2,4,4'',6-Tetrachloro-p-terphenyl	269
368	100	370	96	372	53	366	45	Heptachloronaphthalene (Halowax 1014)	237
368	100	370	98	404	95	402	82	Mirex	343

SPECTRAL PEAK INDEX AT 100 DEG.

m/z	int	m/z	int	m/z	int	m/z	int	ABBREVIATED NAME	PAGE
369	100	371	34	339	24	370	16	3-Chloro-2,6-dinitro-N,N-dipropyl-	197
370	100	250	99	262	90	264	84	Dicofol	319
371	100	127	38	372	9	245	7	Ioxynil	76
374	100	375	19	376	7	344	5	Oryzalin, dimethyl	176
374	100	376	80	372	47	378	36	1,2,3,4,8,9-Hexachlorodibenzofuran	295
378	100	348	43	379	16	349	7	2,6-Dinitro-N,N,N',N'-tetraethyl-4-	201
380	100	378	99	382	66	308	48	Octachlorostyrene	40
380	100	382	80	378	52	384	36	Endrin Aldehyde	330
380	100	382	82	346	78	378	51	Endrin	329
383	100	385	98	387	50	381	44	Kepone	345
386	100	388	71	384	60	422	44	Endosulfan cyclic sulfate	352
388	100	390	98	392	54	318	47	Heptachlor epoxide	337
390	100	391	27	148	21	392	5	Dioctyl phthalate	216
390	100	391	28	148	20	262	9	Di-n-octyl-phthalate	217
390	100	392	82	388	48	394	36	1,2,3,4,7,8-Hexachlorodibenzo-p-dioxin	290
391	100	392	23	361	7	393	3	4-Cyano-2,6-dinitro-N,N,N',N'-	204
394	100	396	71	392	70	390	18	1,2,4,5-Tetrabromobenzene	43
394	100	396	93	398	51	392	46	2,2',3,4,4',5,5'-Heptachlorobiphenyl	250
398	100	396	88	400	66	394	33	Chlorbicyclen	349
400	100	450	62	350	53	512	34	Perfluorokerosene-H	361
404	100	402	87	368	79	370	76	Photomirex (2,8-dihydro mirex)	344
404	100	402	87	406	63	400	36	Octachloronaphthalene	238
404	100	402	90	406	61	400	36	Pentac	342
406	100	408	80	404	53	410	39	Endosulfan II	351
406	100	408	84	404	53	372	42	Endosulfan I	350
408	100	409	94	410	79	411	65	Pentabromophenol	71
408	100	410	97	412	53	406	42	1,2,3,4,6,7,8-Heptachlorodibenzofuran	296
409	100	410	22	321	7	411	3	4-Carbamyl-2,6-dinitro-N,N,N',N'-	205
410	100	408	88	412	65	374	40	gamma-Chlordane	340
410	100	408	89	412	65	406	34	alpha-Chlordane	339
418	100	419	29	420	5	148	5	Dinonyl phthalate	218
424	100	422	91	426	66	352	57	Oxychlordane	338
424	100	426	96	428	52	422	44	1,2,3,4,6,7,8-Heptachlorodibenzo-p-	291
428	100	426	51	430	50	429	19	Bromopropylate	321
430	100	428	90	432	63	426	33	2,2',3,3',4,4',5,5'-Octachlorobiphenyl	251
442	100	443	19	365	5	444	2	Decafluorotriphenylphosphine	357
444	100	442	86	446	64	440	34	Octachlorodibenzofuran	297
444	100	446	75	442	75	448	37	trans-Nonachlor	341
458	100	381	27	459	21	438	7	Decafluorotriphenylphosphine oxide	358
460	100	423	78	425	77	458	76	Octachlorodibenzo-p-dioxin	292
464	100	462	81	466	73	468	36	2,2',3,3',4,4',5,5',6-Nonachloro-	252
486	100	488	94	484	51	490	46	Pentabromotoluene	96
498	100	500	83	496	68	502	47	Decachlorobiphenyl	253
500	100	502	80	498	52	504	35	Dibutyl chlorendate	348
548	100	550	98	546	52	552	51	2,2',4,5,5'-Pentabromobiphenyl	258
552	100	550	78	554	75	472	45	Hexabromobenzene	44
564	100	562	95	81	67	644	65	Hexabromodiphenyl ether	280
628	100	626	79	630	78	624	33	2,2',4,4',5,5'-Hexabromobiphenyl	259
633	100	595	62	634	15	557	15	Perfluorotributylamine	359

SPECTRAL PEAK INDEX AT 250 DEG.

m/z	int	m/z	int	m/z	int	m/z	int	ABBREVIATED NAME	PAGE
35	100	37	29	161	3	159	3	a,2-Dichlorotoluene	89
35	100	37	35	70	25	72	21	1,1,2-Trichloroethane	2
35	100	37	35	193	9	195	8	2,4,5-Trichlorotoluene	91
35	100	37	36	97	1	0	0	1,1,1-Trichloroethane	1
35	100	37	36	97	1	0	0	1,1,1,2-Tetrachloroethane	3
35	100	37	36	159	2	161	1	a,4-Dichlorotoluene	88
35	100	37	36	159	4	161	2	a,a-Dichlorotoluene	87
35	100	37	36	257	7	255	7	2,2',5-Trichlorobiphenyl	245
35	100	37	37	70	4	72	2	Hexachloro-1,3-butadiene	16
35	100	37	37	134	1	132	1	Trichloroethene	10
35	100	37	37	297	16	299	14	2,4'',5-Trichloro-p-terphenyl	268
35	100	37	38	70	22	72	15	Hexachloroethane	6
35	100	37	38	161	1	159	1	a,a,a-Trichlorotoluene	90
35	100	37	38	216	4	215	4	1,2,3,4-Tetrachlorobenzene	37
35	100	37	38	291	36	256	34	2,2',4,5,5'-Pentachlorobiphenyl	247
35	100	37	39	166	13	164	10	Tetrachloroethene	11
35	100	37	39	181	1	179	1	1,2,3-Trichlorobenzene	33
35	100	37	39	215	2	181	2	1,2,4,5-Tetrachlorobenzene	36
35	100	37	40	70	3	72	2	Hexachloropropene	15
35	100	37	40	119	22	153	4	2-Chloroacetophenone	299
35	100	37	40	163	20	161	14	1-Chloronaphthalene	231
35	100	37	40	181	1	179	1	1,3,5-Trichlorobenzene	34
35	100	37	40	181	2	179	2	1,2,4-Trichlorobenzene	35
35	100	37	40	187	3	221	2	2,4'-Dichlorobiphenyl	243
35	100	37	40	232	25	230	23	alpha-Chlordene	333
35	100	37	40	267	6	232	6	gamma-Chlordene	334
35	100	37	41	159	2	254	1	a,2,6-Trichlorotoluene	92
35	100	37	41	243	4	209	3	2,4,5,6-Tetrachloro-m-xylene	97
35	100	37	42	281	23	262	23	DDE-p,p'	306
35	100	37	43	70	11	72	8	Pentachloroethane	5
35	100	37	43	241	4	277	2	1,1-Dichloro-2,2-di-(4-tolyl)-	309
35	100	70	43	37	34	72	29	1,2,3-Trichloropropane	7
35	100	70	60	72	41	37	40	Octachloropropane	9
35	100	70	85	72	60	262	46	1,1-Bis(4-chlorophenyl)-1,2,2,2-	316
35	100	71	57	37	41	73	38	DDD-p,p'	314
35	100	134	35	37	34	242	19	a,a,a,a',a',a'-Hexachloro-m-xylene	100
35	100	174	72	176	57	37	38	a,a,2,6-Tetrachlorotoluene	93
35	100	174	84	176	56	37	39	2,4,6-Trichloroanisole	77
35	100	176	80	205	74	207	49	Chloroneb	80
35	100	197	42	161	34	37	33	1,8-Dichloronaphthalene	232
35	100	223	80	221	67	37	38	4,4'-Dichlorobiphenyl	242
35	100	237	74	239	48	235	44	Hexachlorocyclopentadiene	21
35	100	237	86	330	74	239	61	Isodrin	327
35	100	246	77	248	51	37	39	DDE-o,p'	307
35	100	255	87	257	83	259	34	2,2',5,5'-Tetrachlorobiphenyl	246
35	100	304	88	303	87	305	69	2,2-Bis(4-chlorophenyl)-1,1,1-	311
59	100	60	3	122	2	138	1	2-Nitrophenol, acetyl derivative	105
70	100	72	69	35	62	37	26	1,1,2,2-Tetrachloroethane	4
70	100	232	91	72	70	234	64	2,3-Dichlorohexafluoro-2-butene	14
71	100	73	63	255	51	70	51	BHC, beta isomer	18
71	100	73	64	211	59	213	57	Tetrasul	277
71	100	73	68	35	56	37	20	2,4,4',5-Tetrachlorodiphenyl ether	271

SPECTRAL PEAK INDEX AT 250 DEG.

m/z	int	m/z	int	m/z	int	m/z	int	ABBREVIATED NAME	PAGE
71	100	73	68	35	56	308	37	2',3,4,4'-Tetrachlorodiphenyl ether	272
71	100	73	71	35	60	283	46	DDT-p,p'	312
71	100	73	72	35	48	201	32	2,2',4-Trichlorodiphenyl ether	270
71	100	255	96	73	70	70	62	BHC, alpha isomer	17
79	100	81	96	160	75	162	39	Dibromochloropropane	8
79	100	81	96	469	8	471	6	2,2',4,5,5'-Pentabromobiphenyl	258
79	100	81	97	0	0	0	0	2,4,5-Tribromobiphenyl	256
79	100	81	99	157	1	155	1	Bromobenzene	26
79	100	81	99	488	1	486	1	Pentabromotoluene	96
79	100	81	99	563	78	565	76	Hexabromodiphenyl ether	280
81	100	79	91	314	1	0	0	1,3,5-Tribromobenzene	41
81	100	79	93	411	13	413	7	Bromopropylate	321
81	100	79	94	0	0	0	0	1,2,4-Tribromobenzene	42
81	100	79	95	394	7	396	5	1,2,4,5-Tetrabromobenzene	43
81	100	79	96	35	6	293	2	1-(4-Bromophenyl)-1-phenyl-2,2,2-	317
81	100	79	96	161	30	159	16	Pentabromodiphenyl ether	279
81	100	79	96	161	63	159	32	Tetrabromodiphenyl ether	278
81	100	79	96	331	3	329	3	2,4,6-Tribromoaniline	82
81	100	79	96	628	9	630	6	2,2',4,4',5,5'-Hexabromobiphenyl	259
81	100	79	97	0	0	0	0	1-Bromonaphthalene	230
81	100	79	98	391	5	389	5	2,2',5,5'-Tetrabromobiphenyl	257
81	100	79	99	35	9	37	3	1-(4-Bromophenyl)-2,2,2-trichloro-	323
81	100	219	15	220	1	0	0	4,4'-Dibromobiphenyl	255
102	100	104	69	103	10	115	7	Benzonitrile	28
106	100	93	36	107	14	120	7	Phenol	29
111	100	113	63	35	18	112	10	Chlorobenzene	25
123	100	124	8	107	3	125	1	Nitrobenzene	24
127	100	0	0	0	0	0	0	Iodobenzene	27
137	100	138	9	121	3	106	2	3-Nitrotoluene	118
137	100	138	9	121	4	139	1	2-Nitrotoluene	109
137	100	138	9	121	6	106	2	4-Nitrotoluene	128
138	100	122	46	324	14	139	5	4-Nitroaniline	129
138	100	139	7	122	3	140	1	3-Nitroaniline	119
138	100	139	8	104	6	120	4	2-Nitroaniline	110
138	100	304	10	139	8	122	3	1,2-Bis(4-nitrophenoxy)ethane	286
138	100	304	43	305	8	139	7	1,2-Bis(2-nitrophenoxy)ethane	287
139	100	122	14	140	8	121	8	2-Nitrophenol	108
139	100	140	7	138	2	123	2	3-Nitrophenol	117
139	100	140	7	138	4	123	4	4-Nitrophenol	127
143	100	171	82	155	57	145	37	2-Chloroethyl-p-toluenesulfonate	354
145	100	35	86	147	69	37	28	1,2-Dichlorobenzene	30
145	100	35	93	147	72	37	36	1,3-Dichlorobenzene	31
145	100	147	72	35	56	37	18	1,4-Dichlorobenzene	32
148	100	118	12	149	9	132	6	2-Nitrobenzonitrile	102
148	100	149	8	276	4	291	3	Dinonyl phthalate	218
148	100	149	9	132	8	118	4	4-Nitrobenzonitrile	121
148	100	149	9	132	9	118	3	3-Nitrobenzonitrile	112
148	100	149	10	221	5	279	2	Dibutyl phthalate	215
148	100	149	10	277	3	262	2	Dioctyl phthalate	216
148	100	149	10	277	3	262	2	Di-n-octyl-phthalate	217
148	100	179	14	149	10	195	5	Dimethyl phthalate	213
148	100	193	12	149	9	162	5	Diethyl phthalate	214

SPECTRAL PEAK INDEX AT 250 DEG.

m/z	int	m/z	int	m/z	int	m/z	int	ABBREVIATED NAME	PAGE
151	100	127	73	254	48	278	12	Tetraiodoethene	13
151	100	152	8	121	8	153	1	2-Nitrobenzaldehyde	106
151	100	152	8	135	4	153	1	4-Nitrobenzaldehyde	125
151	100	152	8	135	8	153	1	3-Nitrobenzaldehyde	116
153	100	35	73	155	49	37	27	4-Chloroacetophenone	298
153	100	138	34	137	6	122	4	4-Nitroanisole	126
153	100	154	8	136	6	135	4	4-Methyl-2-nitrophenol	130
153	100	154	9	138	3	46	3	2-Nitroanisole	107
153	100	154	28	167	3	155	3	1-Cyanonaphthalene	228
156	100	157	13	196	1	182	1	1-Naphthaldehyde	229
157	100	159	35	158	7	160	3	1-Chloro-3-nitrobenzene	50
157	100	159	35	158	7	160	3	1-Chloro-4-nitrobenzene	51
157	100	159	35	158	8	160	3	1-Chloro-2-nitrobenzene	49
160	100	158	51	162	48	81	15	Tetrabromoethene	12
160	100	196	91	198	86	161	72	2,3,6-Trichlorophenol	63
163	100	164	9	143	3	182	2	Pentafluoroaniline	86
164	100	165	8	184	4	136	3	Pentafluorophenol	72
166	100	183	78	167	18	153	13	2,4-Dinitroaniline	149
167	100	127	12	168	8	294	1	Iodopentafluorobenzene	48
167	100	168	9	183	6	151	3	2,4-Dinitrophenol,	141
167	100	168	12	154	7	184	4	2,4-Dinitrophenol	143
168	100	138	32	169	8	139	2	1,2-Dinitrobenzene	101
168	100	152	10	138	9	169	8	1,3-Dinitrobenzene	111
168	100	152	10	169	8	138	4	1,4-Dinitrobenzene	120
173	100	174	12	157	7	175	1	1-Nitronaphthalene	227
179	100	181	98	195	89	197	85	1,3,5-Trichloro-2-nitrobenzene	54
179	100	245	45	209	21	247	15	N-(2-Chloroethyl)-2,4-dinitroaniline	145
180	100	181	16	182	1	35	1	2-Chlorobenzophenone	302
181	100	138	53	182	10	139	5	4-Nitrophenol, acetyl derivative	124
181	100	163	54	182	29	162	15	Pentafluorotoluene	94
181	100	182	9	165	3	150	2	4-Nitrobenzoic acid, methyl ester	123
181	100	182	10	138	2	183	1	2-Nitrobenzoic acid, methyl ester	104
181	100	182	10	138	6	165	1	3-Nitrophenol, acetyl derivative	115
181	100	182	10	165	9	183	1	3-Nitrobenzoic acid, methyl ester	114
181	100	182	14	198	12	168	9	DNOC	184
181	100	197	77	183	30	199	26	4-Chloro-3,5-dinitrobenzonitrile	137
181	100	198	87	168	15	182	10	2,6-Dinitro-p-creosol	159
182	100	152	14	183	9	166	4	2,6-Dinitrotoluene	161
182	100	152	30	183	9	166	3	2,3-Dinitrotoluene	164
182	100	183	9	166	9	152	9	2,4-Dinitrotoluene	144
182	100	183	36	184	5	180	5	Benzophenone	300
183	100	164	11	184	7	136	6	Pentafluoroanisole	79
183	100	184	8	153	2	185	1	3,5-Dinitroaniline	136
183	100	184	8	153	5	167	4	2,6-Dinitroaniline	163
183	100	198	15	184	8	167	7	2,4-Dinitroanisole	142
186	100	183	13	187	7	167	7	Hexafluorobenzene	45
191	100	161	11	175	9	192	8	2-Nitrobenzotrifluoride	103
191	100	175	11	192	8	193	1	3-Nitrobenzotrifluoride	113
191	100	192	8	175	2	193	1	4-Nitrobenzotrifluoride	122
192	100	226	75	228	52	194	34	Dichlone	240
193	100	163	35	194	10	147	6	3,5-Dinitrobenzonitrile	131
193	100	194	9	188	1	0	0	Pentafluorobenzonitrile	60

SPECTRAL PEAK INDEX AT 250 DEG.

m/z	int	m/z	int	m/z	int	m/z	int	ABBREVIATED NAME	PAGE
194	100	195	12	196	2	164	2	Dimethyl terephthalate	221
195	100	197	94	199	31	196	10	3,4,5-Trichloroaniline	81
196	100	194	90	195	66	197	63	2,3,5,6-Tetrachlorophenol	68
196	100	194	90	197	75	195	75	2,3,4,5-Tetrachlorophenol	67
196	100	198	91	200	30	197	9	2,3,4-Trichlorophenol	61
196	100	198	93	197	44	195	33	Bromoxynil	75
196	100	198	94	200	31	197	23	2,4,6-Trichlorophenol	65
196	100	198	96	160	47	200	32	2,4,5-Trichlorophenol	64
196	100	198	96	160	86	162	55	3,4,5-Trichlorophenol	66
196	100	198	98	160	44	162	32	2,3,5-Trichlorophenol	62
197	100	198	9	181	5	167	4	2-Amino-4,6-dinitrotoluene	155
197	100	198	9	181	6	180	4	2,4-Dinitro-N-methylaniline	148
197	100	199	97	164	68	166	40	2,2,4,5-Tetrachlorocyclopentene-	23
198	100	180	68	244	20	199	14	2,2'-Dinitrobiphenyl	260
198	100	199	9	182	7	168	3	3,5-Dinitrobenzyl alcohol	135
200	100	217	57	202	35	219	17	6-Chloro-2,4-dinitroaniline	153
202	100	203	21	216	5	204	2	Fluoranthene	224
202	100	204	31	203	8	205	2	Chloropentafluorobenzene	46
202	100	204	33	172	29	174	10	1-Chloro-2,4-dinitrobenzene	140
204	100	217	45	170	27	205	14	N-(2-Chloroethyl)-4-cyano-2,6-	212
206	100	208	64	189	21	191	14	4,5-Dichloro-2-nitroaniline	150
206	100	208	65	210	10	189	10	2,4-Dichloro-6-nitroaniline	151
206	100	208	65	210	10	207	8	2,6-Dichloro-4-nitroaniline	152
208	100	210	95	212	22	209	18	1-(4-Chlorophenyl)-2,2,2-trifluoro-	322
209	100	226	38	210	15	208	11	2,4-Dinitro-6-isopropylphenol	185
211	100	181	12	212	9	195	8	3,5-Dinitrobenzamide	133
211	100	212	11	181	9	195	2	2,6-Dinitro-N-ethylaniline	162
212	100	214	69	176	24	213	17	2,4-Dichloronapthol	241
215	100	213	79	217	48	231	23	2,3,5,6-Tetrachloronitrobenzene	56
215	100	217	68	219	12	216	9	2,4-Dichloro-3,5-dinitrobenzonitrile	165
216	100	218	33	217	19	219	5	4-Chlorobenzophenone	301
216	100	252	81	254	53	218	36	2,4,4'-Trichloro-2'-hydroxydiphenyl	281
217	100	219	37	218	16	220	5	Diclofop-methyl	285
218	100	216	77	220	48	222	11	Hexachloro-3-cyclopentenone	22
218	100	216	80	220	48	383	37	Kepone	345
218	100	219	15	220	1	0	0	4,4'-Difluorobenzophenone	304
218	100	219	43	216	23	214	21	4,4'-Difluorobenzhydrol	325
221	100	35	82	223	75	37	26	2,2'-Dichlorobiphenyl	244
221	100	148	19	222	14	223	2	Butyl benzyl phthalate	220
223	100	224	12	222	7	239	3	Dinoseb Acetate	187
223	100	240	55	224	18	222	11	Dinoseb	186
224	100	226	97	228	32	242	9	1,3-Dinitro-2,4,5-trichlorobenzene	168
225	100	195	45	227	33	197	14	5-Chloro-2-nitrobenzotrifluoride	158
225	100	227	33	226	8	209	8	2-Chloro-5-nitrobenzotrifluoride	156
225	100	227	34	195	24	197	8	4-Chloro-3-nitrobenzotrifluoride	157
226	100	196	15	227	10	210	10	3,5-Dinitrobenzoic acid,	134
227	100	225	96	195	54	197	47	1,2,4-Trichloro-5-nitrobenzene	53
227	100	225	99	195	69	197	67	1,2,3-Trichloro-4-nitrobenzene	52
230	100	231	24	232	2	258	1	Diphenylfulvene	223
230	100	231	83	228	73	229	66	Pentachlorophenol	70
230	100	232	95	234	32	231	15	Trichloronaphthalene (Halowax 1014)	233
231	100	35	83	229	76	233	45	2,3,5,6-Tetrachloroaniline	84

SPECTRAL PEAK INDEX AT 250 DEG.

m/z	int	m/z	int	m/z	int	m/z	int	ABBREVIATED NAME	PAGE
231	100	229	77	233	48	235	10	2,3,4,5-Tetrachloroaniline	83
231	100	261	79	229	75	259	60	2,3,4,5-Tetrachloronitrobenzene	55
232	100	234	91	266	79	268	50	Chlordene	332
232	100	322	41	292	33	305	25	Dinitramine	193
233	100	235	67	249	11	237	11	2,4-Dichloro-3,5-dinitrobenzamide	167
235	100	236	9	252	3	222	3	2,6-Dinitro-4-(trifluoromethyl)phenol	160
236	100	220	10	206	10	237	9	3,5-Dinitrobenzotrifluoride	132
237	100	35	64	235	62	239	59	Chlorbicyclen	349
237	100	35	89	239	71	235	71	Aldrin	326
237	100	209	60	250	42	233	34	N-(2-Chloroethyl)-2,6-dinitro-4-	211
237	100	221	30	222	20	267	13	4-Carbamyl-2,6-dinitro-N,N,N',N'-	200
237	100	239	67	235	65	35	47	Dieldrin	328
237	100	282	94	318	86	280	72	Heptachlor epoxide	337
239	100	240	12	223	9	165	7	N,N-Diethyl-2,4-dinitroaniline	146
240	100	270	93	224	65	242	35	4-Chloro-3,5-dinitrobenzotrifluoride	138
242	100	240	65	81	63	79	62	4-Bromo-2,5-dichlorophenol	74
242	100	240	72	35	67	406	55	Endosulfan II	351
242	100	240	77	244	44	35	44	Endosulfan I	350
242	100	240	80	244	50	243	13	a,a',2,3,5,6-Hexachloro-p-xylene	98
242	100	240	85	244	50	246	15	a,a,a,a',a',a'-Hexachloro-p-xylene	99
244	100	265	81	242	78	280	53	Pentachloroanisole	78
246	100	244	79	247	60	248	58	Pentachlorothiophenol	73
246	100	244	99	263	56	261	55	6-Bromo-2,4-dinitroaniline	154
246	100	248	69	283	26	281	26	DDT-o,p'	313
246	100	248	84	35	49	250	25	DDD-o,p'	315
246	100	248	96	81	18	79	18	Bromopentafluorobenzene	47
247	100	219	65	260	37	261	23	N-(2-Chloroethyl)-2,6-dinitro-4-	207
247	100	219	93	277	47	213	45	N-(2-Bromoethyl)-2,6-dinitro-4-	208
247	100	249	62	35	48	211	27	DDD-p,p'olefin	308
249	100	219	19	279	14	250	14	4-Cyano-2,6-dinitro-N,N,N',N'-	199
249	100	251	65	247	61	253	19	Pentachloronitrobenzene	57
250	100	252	65	248	60	254	22	Pentachlorobenzene	38
250	100	252	66	251	15	254	12	4,4'-Dichlorobenzophenone	303
250	100	252	69	254	18	251	16	4,4'-Dichlorobenzhydrol	324
252	100	253	24	254	3	266	2	Benzo[a]pyrene	225
255	100	71	95	257	75	73	68	BHC, gamma isomer	20
255	100	257	66	253	59	71	47	BHC, delta isomer	19
258	100	260	65	262	11	259	8	2,4-Dichloro-3,5-dinitrobenzo-	166
260	100	230	40	262	36	232	13	4-Chloro-3,5-dinitrobenzoic acid,	139
261	100	335	65	305	41	245	21	4-Cyano-2,6-dinitro-N,N,N',N'-	202
262	100	264	63	250	54	252	36	Chlorobenzilate	320
262	100	264	69	250	38	252	24	Dicofol	319
264	100	262	67	266	62	194	22	Pentachlorotoluene	95
264	100	266	65	284	26	265	21	4,4'-Dichloro-a-(trifluoromethyl)-	318
265	100	263	63	267	62	229	55	Pentachloroaniline	85
265	100	263	65	267	64	269	21	Decachlorodiphenyl ether	276
266	100	264	73	268	49	270	11	Chlorothalonil	59
266	100	264	74	268	48	267	13	Tetrachloronaphthalene (Halowax 1014)	234
266	100	264	76	410	70	408	62	gamma-Chlordane	340
266	100	264	78	268	53	237	36	alpha-Chlordane	339
266	100	264	79	300	55	268	50	Heptachlor	336
266	100	282	87	264	77	280	59	1-Hydroxychlordene	335

SPECTRAL PEAK INDEX AT 250 DEG.

m/z	int	m/z	int	m/z	int	m/z	int	ABBREVIATED NAME	PAGE
267	100	268	15	251	7	237	5	2,4-Dinitro-N,N-dipropylaniline	147
271	100	272	16	176	10	273	9	2,2-Bis(4-fluorophenyl)-1,1,1-	310
272	100	237	66	270	60	238	60	Endrin	329
272	100	270	86	308	52	274	44	Endrin Aldehyde	330
272	100	270	89	274	40	307	34	2',3',4',5,5'-Pentachloro-2-biphenyol	267
272	100	270	93	307	68	305	51	2',3,3',4',5'-Pentachloro-2-biphenyol	266
272	100	270	99	274	33	308	28	2',3',4',5'-Tetrachloro-4-biphenyol	263
272	100	273	12	274	1	267	1	Octafluoronaphthalene	239
272	100	273	14	254	3	273	1	2,3,4,5,6-Pentafluorobenzophenone	305
272	100	273	17	237	4	274	2	2,2'-Dinitrobibenzyl	261
272	100	273	17	256	3	274	2	4,4'-Dinitrobibenzyl	262
275	100	277	67	273	63	279	22	Pentachlorobenzonitrile	58
276	100	277	26	290	11	278	5	Benzo[ghi]perylene	226
276	100	278	66	277	19	280	11	4,4'-Dichlorochalcone	355
278	100	295	24	279	22	296	6	Dinocap 5	192
278	100	295	30	279	27	296	7	Dinocap 3	190
279	100	249	22	280	10	263	3	2,6-Dinitro-N-ethyl-4-	180
279	100	249	89	280	11	250	9	N,N-Dimethyl-2,6-dinitro-4-	173
279	100	280	16	306	12	263	12	4-Carbamyl-2,6-dinitro-N,N,N',N'-	203
279	100	295	54	280	18	85	10	Dinocap 1	188
279	100	295	57	280	18	296	10	Dinocap 2	189
279	100	295	64	278	50	280	18	Dinocap 4	191
281	100	282	16	251	4	283	2	Pendimethalin	182
283	100	267	66	313	64	237	55	3-Chloro-N,N-dimethyl-2,6-dinitro-	195
283	100	285	65	284	17	287	11	Nitrofen	282
284	100	286	85	282	54	288	36	Hexachlorobenzene	39
286	100	288	99	290	32	287	14	3,3',5,5'-Tetrachloro-4,4'-biphenyldiol	265
287	100	285	91	322	83	320	75	1,2,3,4-Tetrachlorodibenzo-p-dioxin	288
289	100	203	92	319	82	261	61	Fluchloralin	206
292	100	262	51	293	16	263	8	4-Cyano-2,6-dinitro-N,N-dipropyl-	179
292	100	322	74	262	33	293	13	2,6-Dinitro-N,N,N',N'-tetramethyl-4-	198
293	100	263	15	294	13	277	4	2,6-Dinitro-N-propyl-4-	181
295	100	201	23	248	15	265	13	2,6-Dinitro-N-(2-hydroxyethyl)-4-	209
295	100	296	18	265	3	297	2	Butralin	183
296	100	295	36	361	31	332	17	Oxyfluorfen	283
299	100	297	73	301	54	300	14	1,4,5,6,7,7-Hexachloro-5-norbornene-	346
300	100	302	66	298	60	444	46	trans-Nonachlor	341
300	100	302	71	298	62	304	22	Pentachloronaphthalene (Halowax 1014)	235
303	100	391	38	361	26	287	20	4-Cyano-2,6-dinitro-N,N,N',N'-	204
304	100	378	73	348	23	305	15	2,6-Dinitro-N,N,N',N'-tetraethyl-4-	201
306	100	304	78	308	52	307	15	1,2,7,8-Tetrachlorodibenzofuran	293
306	100	342	97	304	81	344	62	2,2',4,4',5-Pentachlorodiphenyl ether	274
307	100	277	35	308	13	278	5	N,N-Diethyl-2,6-dinitro-4-	172
308	100	272	62	310	60	306	60	Endrin Ketone	331
308	100	306	72	310	50	272	39	3,3',4,4'-Tetrachlorodiphenyl ether	273
308	100	306	77	310	49	309	14	2',3',4',5'-Tetrachloro-3-biphenyol	264
308	100	310	84	274	62	306	53	Octachlorostyrene	40
309	100	310	19	311	3	279	3	Isopropalin	174
316	100	318	64	314	64	352	58	Oxychlordane	338
317	100	304	29	318	21	334	20	Dinitramine, acetyl derivative	194
318	100	148	38	319	23	93	6	Diphenyl phthalate	219
320	100	318	73	322	48	321	16	3,3',4,4'-Tetrachloroazobenzene	356

SPECTRAL PEAK INDEX AT 250 DEG.

m/z	int	m/z	int	m/z	int	m/z	int	ABBREVIATED NAME	PAGE
320	100	318	92	322	34	245	24	Tetradifon	353
321	100	322	18	305	10	364	8	4-Carbamyl-2,6-dinitro-N,N,N',N'-	205
325	100	326	16	295	11	293	4	2,6-Dinitro-N,N-dipropyl-4-	178
326	100	328	64	324	63	330	20	2,3,4,5,6-Pentachlorobiphenyl	248
327	100	329	35	328	14	261	11	N-(3-Chloropropyl)-2,6-dinitro-4-	210
330	100	332	99	81	95	79	93	2,4,6-Tribromophenol	69
332	100	330	81	334	49	333	13	Dacthal	222
333	100	297	96	295	73	335	66	1,4,5,6,7,7-Hexachlorobicyclo[2.2.1]-	347
333	100	331	96	35	78	335	36	2,4,4'',6-Tetrachloro-p-terphenyl	269
334	100	335	12	336	10	315	1	Decafluorobiphenyl	254
334	100	336	85	332	53	338	37	Hexachloronaphthalene (Halowax 1014)	236
335	100	305	19	336	16	303	4	Benefin	170
335	100	305	20	336	16	319	4	Trifluralin	169
340	100	342	66	338	58	344	22	1,2,3,8,9-Pentachlorodibenzofuran	294
341	100	311	82	295	67	343	33	3-Chloro-N,N-diethyl-2,6-dinitro-	196
341	100	343	68	342	19	345	12	Bifenox	284
342	100	344	64	340	62	306	37	3,3',4,4',5-Pentachlorodiphenyl ether	275
345	100	315	27	346	17	347	7	Nitralin	177
346	100	347	16	316	10	255	9	Oryzalin	175
347	100	317	19	348	18	331	4	Profluralin	171
350	100	300	9	351	8	400	5	Perfluorokerosene-L	360
350	100	400	56	281	42	331	27	Perfluorokerosene-H	361
355	100	357	65	353	62	390	57	1,2,3,4,7,8-Hexachlorodibenzo-p-dioxin	290
356	100	358	64	354	63	321	58	1,2,3,4,7-Pentachlorodibenzo-p-dioxin	289
360	100	362	82	358	51	364	33	2,2',4,4',5,5'-Hexachlorobiphenyl	249
369	100	339	62	323	43	371	33	3-Chloro-2,6-dinitro-N,N-dipropyl-	197
370	100	368	97	372	54	366	45	Heptachloronaphthalene (Halowax 1014)	237
371	100	127	87	244	53	243	22	Ioxynil	76
374	100	344	20	375	19	376	8	Oryzalin, dimethyl	176
374	100	376	82	372	53	378	36	1,2,3,4,8,9-Hexachlorodibenzofuran	295
375	100	299	81	430	75	373	73	Dibutyl chlorendate	348
386	100	97	93	388	72	384	60	Endosulfan cyclic sulfate	352
389	100	391	84	387	53	393	34	1,2,3,4,6,7,8-Heptachlorodibenzo-p-	291
394	100	396	95	398	50	392	45	2,2',3,4,4',5,5'-Heptachlorobiphenyl	250
404	100	402	85	406	65	370	50	Photomirex (2,8-dihydro mirex)	344
404	100	402	86	370	86	368	86	Mirex	343
404	100	402	89	300	65	406	64	Pentac	342
404	100	402	89	406	63	400	35	Octachloronaphthalene	238
408	100	410	69	409	67	406	66	Pentabromophenol	71
408	100	410	97	412	53	406	44	1,2,3,4,6,7,8-Heptachlorodibenzofuran	296
425	100	423	99	427	51	421	42	Octachlorodibenzo-p-dioxin	292
430	100	428	89	432	65	426	34	2,2',3,3',4,4',5,5'-Octachlorobiphenyl	251
442	100	365	28	167	24	443	20	Decafluorotriphenylphosphine	357
444	100	442	89	446	65	440	36	Octachlorodibenzofuran	297
458	100	381	95	438	35	459	21	Decafluorotriphenylphosphine oxide	358
464	100	462	79	466	74	468	36	2,2',3,3',4,4',5,5',6-Nonachloro-	252
498	100	500	86	496	67	502	50	Decachlorobiphenyl	253
552	100	554	77	550	73	81	58	Hexabromobenzene	44
633	100	595	57	452	14	634	13	Perfluorotributylamine	359